Northwest Michigan Engine and Thresher Club

Buckley Old Engine Show

A Fifty Year History

AF698972

Arbutus Press
Traverse City, Michigan

Buckley Old Engine Show: A Fifty Year History © 2017, Northwest Michigan Engine and Thresher Club
ISBN 978-1-933926-56-8

All rights reserved. No part of this book may be reproduced or transmitted in any form by any means, electronic, or mechanical, including photocopying and recording, or by any information storage and retrieval system, except for brief excerpts or by permission of the publisher.

Printed in the United States of America

Arbutus Press
Traverse City, Michigan
editor@arbutuspress.com
www.Arbutuspress.com

image credit title page : 1916 Victor tractor made by the Victor Traction Gear Company. The Victor belongs to the club and is driven by Mike Hartwig in the photo.

image credit page 4, John Russell: Engine #207 giving children rides on a snowy fall Education Day.

ACKNOWLEGEMENTS AND PREFACE

The book committee and its contributors

Jim Luper	Karen Youker
Larry Clous	Steve Scott
Frank D. Holzschu	Sara Fenten
Ham Hobson	John Russell
Barb Hobson	Jan Clous
Justin Gelinas	Ron Decker
Lori Gelinas	Bob Hartwig

The Buckley Old Engine Club is blessed to have had a group of founders who saw a need to keep the older methods of farming in practice to show the basis of how we modernized to where we are today. This included all aspects of farm life from the fields, to the barns, to all of the sheds and coops, to the power sources and transportation, to the home, etc. It became much more than a passion, it became a need to show the history of development and a need to keep all of this alive and not just a chapter in an old book or album.

They not only loved what they were about, but they instilled this love and dedication to everyone who took an interest in their "HOBBY". This was a family affair for them and that allowed the experience to spill over to everyone in the family from young to old and from one generation to another. Their spirit and drive were the stepping stones for hundreds to follow and today we are continuing their mission as we have members from little children to young adults and families of all ages that work together to continue to be dedicated and enthusiastic supporters of the club's continued mission to offer an enjoyable experience for all. They truly believe in our motto of PUTTING THE PAST IN MOTION.

As more and more of the common farm practices of the late 1900's give way to the modern trend of larger farms with modern technology, our club members also realize that many of our younger visitors and members are very interested in that gap that has been established between the original farm equipment and agricultural practices of today. Many collectors are restoring the machinery and remembering the farm practices of their youth. Our vision not only includes the old steamers and early gas tractors and all that they represent, but also that equipment that was used in farming as the machinery quickly became bigger and better in the 1960-1990 era when family farms with 10-20 acre fields increased in size thanks to bigger and better equipment. With the addition of two large pieces of land and the club being blessed with many younger members, our vision for the future continues to expand by leaps and bounds as each new idea is presented. I am very proud of our members and their goal of keeping the show new and updated while making it a true family experience for all visitors. The club is in good hands and has excellent new leadership and the goals for the future are truly exciting.

Ron Decker, President of the Northwest Michigan Engine and Thresher Club, 2017

THE START OF A TRACTOR SHOW

Leonard Clous was the sixth oldest of fifteen children born to Frank Clous. Raymond, his brother was two years younger. They grew up on their father's farm in Hannah, Michigan where they both went to school at Hannah Saint Mary's and both graduated the eighth grade. On May 17, 1938, Leonard married Geneva Pelton. They purchased a farm on County Line Road, just west of Buckley, in Wexford County and raised four children. They raised corn, wheat, oats, hay, buckwheat, rye, potatoes and beef cattle on the farm. Three years later on February 11, 1941 Leonard's brother Ray married Irma Worm and they bought a farm just east of Leonard's but on the opposite side of the road in Grand Traverse County. Like Leonard, Ray raised the same crops but Ray also had milk cows and hogs. They each owned their own tractors but as families will do sometimes, they shared the major tilling and harvest equipment and when major projects came up, they worked together to complete them. Haying time was always one of those times. In 1964 I was working a field on Uncle Leonard's farm when my dad and Uncle Leonard came to get me to go with them to see Roy Howes, a vegetable crop farmer in Copemish, Michigan. We planned to talk about getting into growing green snap beans and Roy helped us set a meeting with the managers of Stokely-Van Camps a cannery out of Scottville, Michigan. The next spring in 1965 Leonard and Ray raised the first string beans under contract in the Buckley area. String beans were a huge boost to the farm's income. We needed newer and larger tractors and equipment, upgraded vehicles and additional land purchases. Uncle Leonard was always into antique tractors and equipment. By then he already owned the 1913 Bull tractor that he purchased from a neighbor. The Bull tractor was his inspiration to start his own old tractor field day. This took place in the field next to his house and he would invite neighbors and friends who owned antique tractors to come and show them off. There would be a yard full of men, women and kids watching. Even passers-by would stop to watch.

Once, I was asked to ride along with Uncle Leonard and my dad when Uncle Leonard was interested in a 1928 Baker 25-50hp tractor (or also known as a 43-67hp after the 1929 Nebraska Test), that he found in Oil City, Michigan. The three of us hauled it home on dad's 1958 Ford flatbed truck. Leonard broke his arm trying to start it. The Baker tractor was placed in the southeast corner of the Buckley show grounds with a sign advertising the Engine Show. In 1968 Leonard purchased a 1925 Model "L" Rumley Oil Pull 15-25hp found in a junk yard near Ludington, Michigan. Once again, the three of us went to haul it back. The Oil Pull tractor made it to the first show at the Buckley show grounds. The clutch never did work right so sometimes it made the parade on its own and sometimes it needed towed.

The mini shows worked so well that Leonard got together with a group of guys from different counties to see about organizing an actual show on a bigger scale. That was the beginning of the Northwest Michigan Engine and Thresher Club "DBA" (The Buckley Old Engine Show).

In August of 1967, on the Joe Rebman farm, we held the first Old Engine Show. We gathered around our TV sets and in colorful black and white, and watched Leonard tell everyone to gather up family, friends and neighbors and bring them to the first Old Engine Show The first show had a small turnout it was still big enough to convince the organizers they needed to push forward and continue doing shows. They organized a non-profit club with founding members: Leonard Clous, Raymond Clous, Joe Rebman, William Zue, Levi (Sam) Zue, Clarence Bronson, Clarence Coff, Maurice Hulett, Carl Kreiser, Dan Bantlet, Ervin Blough, Dan Blough, Dale Gassman, C. Babcock, Clark Alexander, Keith Duff and Basil Moored. *Larry Clous*, Buckley, MI, 2017

LEADERS OF THE CLUB

PRESIDENT

1967 - 1988	Leonard Clous
1989 - 1992	Tom Graham
1993 - present	Ron Decker

PRESIDENT EMERITUS

1989 - 2008	Leonard Clous

PAST PRESIDENT

1993 -present	Tom Graham

VICE PRESIDENT

???? - 1982	Joe Smith (no record when Joe became VP or previous VP's)
1983 - 1983	Emerald Magee
1984 - 1985	Jim Puff
1986 - 1989	Tom Graham
1990 - 1992	Fred Mida
1993 - 1997	Tom Graham
1998 - 2002	Richard Nurmi
2003 - 2012	Nate Boynton
2013 - present	Bill Youker

SECRETARY

1968 - 1982?	Joe Rebman
1983 - 1988	Tom Small
1988 - 1992	Margaret Kreiser (took over after Tom's passing)
1993 - 2002	Carol Clous
2003 - 2003	Bev Shillington
2004 - 2014	Frank D. Holzschu
2015 - present	Mike Woodward

TREASURER

1968 - 1980?	Joe Rebman
1981? - 1985	Don Winchell
1985 - 1992	Ron Decker (took over 8/4/95 after Don resigned)
1993 - 2002	Fred Mida
2003 - 2004	Dan Grabill
2005 - 2013	Gloria Crofoot
2014 - present	Jan Clous

FINANCE SECRETARY (in 2003 the club had a Financial Secretary and a Treasurer)

2003 - 2007	Carol Clous & Chris Decker
2008 - 2015	Carol Clous (Carol passed away in 2015)
2015 - 2015	Chris Decker (from Carol's passing until election)
2016 - present	Barb Hobson

FINANCIAL ADVISOR

2003 – 2010	Fred Mida

RECORDING SECRETARY

2000 - 2000	Mary Kay Grubb
2001 - 2002	Bev Shillington

BOARD OF DIRECTORS / TRUSTEES

Two of the Trustees during the 1970's were Raymond Clous and Morris Hulett.

1981 - 1984	Art Clous
1982 - 1996	Herb Kellogg
1985 - 2000	Carl Kreiser
1986 - 1989	Joe Rebman
1990 - 1992	Gib McAllister
1994 - 1999 / 2003 - 2008 / 2012 (7 months)* / 2013 - 2016	Jim Luper
1998 - 2000 / 2004 - 2009	Jim Shillington
2002 - 2004 / 2011 - 2013	Ron Bancroft
2000 - 2005	Duane Gulliver
2003 - 2008	Richard Nurmi
2005 - 2012	Steve Scott
2006 - 2009	Bev Shillington
2009 - 2011	Jim Gunnett
2010 - 2012	Bill Youker
2011 - present	Brandon Jenkins
2012 (7 months) *	Ron Clement
2012 - present	Carl Van Wylen
2012 - present	Larry Clous
2013 - present	Nate Boynton
2014 - 2016	Norm Schutte
2015 - present	Frank D. Holzschu
2016 - present	Todd Benson

* The club changed the terms of the trustees to two trustees elected each year for a three-year term. Adjustments in two trustee terms were made to make this work. In 2002 the club went from three Trustees to six. Club records have been lost or destroyed so most of the information has come from memory, stories told from members no longer with us, stories from members that have been with us from almost the beginning, relatives of the earlier members, pictures and articles. It has been a pleasure bringing you a little insight on what took place as my uncle with the help of my dad and their friends were able to create the greatest Old Engine show on earth.

ABOUT THE FOUNDING MEMBERS

Leonard Clous: Founding member, the man who took his vision of a steam engine show and with a select few made it what it is today. Leonard was the Clubs first President and guided us for over 20 years, he was one of the members that signed the original Articles of Incorporation for the Buckley Club.

LEONARD CLOUS – Lived three miles west northwest of the Buckley Engine show grounds on County Line Road and was the first club president. He was the driving force to where the club is today and was the owner of the club mascot, the 1913 Bull tractor. He was responsible for acquiring most of the club owned land and the club owned tractors like the Flour City. Leonard was one of three signers for the original Articles of Incorporation for the Buckley Club. Leonard passed away in November, 2008 at age 93.

RAYMOND CLOUS – Lived two and a half miles west northwest of the Buckley Engine show grounds on County Line Road. Later he became a club trustee and was known for telling them the buildings they were about to build were not big enough. Then he would smile a year or two later when the additions were built. Ray furnished the loader for lifting the heavy items. His loader hauled logs to and the lumber away from the sawmill from the very first show in Copemish, until 1998 when he found a loader for the club to purchase. This gave the club full time access to a loader on the grounds. Ray was born on the twenty fourth of December and would tell people that he was older than Jesus, because Jesus was born on the twenty fifth and he was born the day before. Ray passed away in September, 2015 at age 97.

Raymond Clous: Founding member shown here operating one of the loaders he furnished for the Club to lift the heavy equipment and haul the logs and lumber for the sawmill.

JOE REBMAN – Lived two miles east of Copemish on Faylor Road and was the first club secretary, sort of. When they started the legal work Keith Duff's secretary volunteered or was coaxed into helping out. Keith owned a real-estate office so his secretary knew her way around the legal system and could dig into areas that helped the boy's achieve their goals. After they had things well organized and her services were no longer needed, Joe was voted in as Secretary/Treasurer. Without a clubhouse in the beginning, the members would volunteer their homes as a meeting place. One of the meetings at our house was the one where Joe was elected to office. The meeting took place at our dining room table. I was 16 years old and sitting at the kitchen island. Leonard handed the pad and pencil to Joe. I was shocked. There Joe sat. His

Joe Rebman: Founding member shown here with his 1917 Port Huron 24-75hp. Joe was the Clubs first secretary / treasurer.

hair was black and ruffled, he had a two or three-day old beard, wore his well-used cap, bib overalls, flannel shirt and work boots. I didn't think he looked like he had enough sense to come in out of the rain let alone be asked to take over as club Secretary/Treasurer. Well like they say don't judge a book by its cover. Joe turned out to be a very intelligent man. The first two shows were held on Joe's farm. Mainly because he owned a 1917 Port Huron 24-75 hp steam traction engine and only had a little over a mile to get it to the site. Joe passed away in October, 1989 at age 70.

WILLIAM ZUE – Lived in the 500 block of Twelfth Street in Traverse City, Michigan. Bill owned Zue Collision Service from the 1940's to the mid 1960's. Located at 2801 U.S. 31 south (just north of where Red Lobster is located now). He was an auto body man and mechanic. Bill was very talented in his field. At the early engine show's, you could have seen Bill driving his early fifties Chevy car that he re-fabricated the rear end into a pickup bed. He actually built a Chevy El-Camino before the first El-Camino's came out in 1959. He also owned a 1917 Port Huron 19-65hp steam traction engine but never got the chance to completely restore it. He sold it to his brother Sam who finished the restoration. Bill was one of three signers for the original Articles of Incorporation for the Buckley Club. Bill passed away in November of 1969 shortly after the first show that was held on the new Buckley Show grounds. Bill was 56 years old when he died.

William Zue: Founding member operating the 1917 Port Huron 19-65hp that he sold to his brother Sam. William was one of the members that signed the original Articles of Incorporation for the Buckley Club.

Sam Zue: Founding member. Sam was a whirlwind of energy, his expertise is shown in many areas throughout the club grounds and as the head sawyer in the sawmill there was none better.

LEVI (Sam) ZUE – Lived two miles north of the Buckley Engine Show Grounds. Sam was a hard-working force. Sam designed and helped build most of the original club buildings including the clubhouse. Sam was also handy with a wrench where ever one was needed whether rebuilding his own steam engines, helping others or anything needing repair on the grounds. He designed and laid-out the existing sawmill and with help, built the low roof covering it, keeping it in line with sawmills back in the day. He rebuilt and directed installation of the Skinner engine and boiler. Sam was responsible for locating, rebuilding and installing the Anderson Hot-Head engine. Sam loved being the sawyer in the club's sawmill. He didn't just put on a show, it was a production mill and that is how he ran it. Just ask anyone that may have worked on the discharge end of the mill and carried the lumber away to stack it. Leonard asked if I could help stack lumber at the mill one afternoon when they were short of help. Short of help? It was two guys Carl Zue (Sam's brother) and myself. Carl ran the edger. Sam saw his skeleton crew and thought this would be a good time for the oak. The big oak, the ones 2x12s come out of. I must have done alright because at the end of the shift Sam looked over to see I had kept up pretty well and gave a little nod and a smile. Sam never said much. If he looked at you and smiled, you knew you did good. I will never forget the joy of watching him drive his 1917 Port Huron 19-65hp steam traction engine down Wilson road and N. Eleven Road on his way to the Buckley show grounds. Sam's profession was a bridge builder mostly hammering the red hot rivets. He was one of the men that helped built the Mackinac Bridge. The next time you have the opportunity to cross the "Mighty Mac" bridge take a minute to look at it and imagine what it must have been like in the fifties to work on such a structure and remember it wasn't just built in the nice warm summer days. Sam passed away in June, 1994 at age 86. Sam was a whirlwind of energy. His expertise is shown in many areas throughout the club grounds and as the head sawyer in the sawmill there was none better.

CLARENCE BRONSON – What I believe is correct, Clarence Bronson lived in the Mayfield, Michigan area and would have been 80 years old when he became a founding member. He was 88 years old when he died in May, 1976. Unfortunately, I have no other information on Mr. Bronson.

CLARENCE COFF – It is possible that Clarence Coff lived in Muskegon, Michigan and worked in a foundry as a molder. Clarence would have been 60 years old when he became a founding member. Unfortunately, I was unable to find any other information on Mr. Coff.

Morris Hullett: Founding member shown here with his pride and joy, a 15hp screen cooled Victor engine built by International Harvester.

MAURICE (MORRIS) HULETT – Lived on a small farm six miles east of the Buckley Engine Show Grounds, on Blackman road. Morris drove a red 1960's Ford Econoline pickup that he used to haul his hit and miss engines to the show. His pride and joy was a 15 horse power screen cooled Victor engine built by International Harvester. Morris ran the veneer mill with his brother-in-law the first few years after the veneer mill came to the show grounds. Morris was a genuine contributor to the show for many years. In 1983 some of his family sheared sheep with a hand powered clippers while riding in the parade. Morris referred to himself as blind in one eye and couldn't see out the other because Morris had a glass eye. If you were a kid and you were lucky (or maybe not so lucky) when you visited him, he would pop his glass eye out and place it in your lap and say "I am keeping my eye on you". Morris passed away in October, 1983 at age 72.

CARL KREISER – Lived three and one half miles north of the Buckley Engines show grounds. He had a machine shop that was the envy of anyone visiting it. Carl had enough lathes to keep a half dozen projects going at one time and other tools to make or rebuild any part imaginable. His shop was full of parts, a lot of them from machines the average person had never heard about. Carl would walk through a flea market looking for antique automobiles and tractor parts. He could point at a table full of parts and tell you the make, model and usually the year of the vehicle that they came off from. Carl found and restored many antique tractors. A few of them are now owned by the Buckley club. He also restored many cars and trucks. When finished they looked like they had just rolled off the assembly line or better. When Carl passed on it was a tremendous loss for the club. Carl passed away in February, 2009 at age 79.

Carl Kreiser: Shown here on his 1921 LaFrance Brockway Torpedo Fire Truck. Carl was the go to man for any information needed for antique automobiles, tractors or trucks, were to find them, were to find their parts, how to make their parts that were no longer available and how to rebuild them.

DAN BANTLET – It is possible that Dan Bantlet lived in Baldwin, Michigan and would have been 62 years old when he became a founding member. Unfortunately, I have no other information on Mr. Bantlet.

ERVIN & DAN BLOUGH
Lived in the Honor, Michigan area and were true engineers. They owned, operated and had been around steam engines their entire lives. They knew how to repair and operate steam engines like the average person knows how to eat with a fork. They also owned and operated their own portable sawmill. During the early years of the club they brought their sawmill to the shows to demonstrate sawing logs into lumber. At the first show in Copemish they got their truck stuck in the sand while they were unloading their Case 75hp

Ervin and Dan Blough: Founding members shown here on their Case 75hp steam traction engine. Ervin and Dan furnished their own portable sawmill and demonstrated it at the first show in Copemish and many shows after until the club acquired its own.

steam traction engine. They didn't have enough room to hook the engine to the front of the truck, so they decided to hook the engine to the side of the trailer. When the engine took off, it tipped the trailer almost on its side before the truck came out unintentionally showing the awesome power of the engine. Ervin and Dan furnished their own portable sawmill until Vern Rennells donated his own sawmill to the club as a permanent addition to the grounds. Ervin passed away in November, 1994 at age 83. Dan passed away in January, 2003 at age 71.

DALE GASSMAN – Lived on Avenue "C" in Traverse City, Michigan. Dale had an old carriage house/coach house behind his home that he turned into his workshop. A carriage house was where you kept your horse and carriage that you used for transportation before automobiles were readily available or affordable. Most carriage houses had a second story or loft were they stored the hay for the horse. Dale worked on his hit and miss engines in his workshop and brought them to the Engine show to display. Dale also had a Buffalo Pitts steam traction engine but never got it running to bring to the show. Dale passed away in September, 1973 at age 74.

C. BABCOCK – Unfortunately I have no information on Mr. Babcock. We do know he was asked to be one of the founding members to what was to become one of the greatest Steam Engine Clubs in the world. It was a great honor for you were recognized as having the knowledge and skills to be a great asset to the Club.

CLARK ALEXANDER – From what I believe is correct. Clark Alexander lived in Osceola county in the Evert, Michigan area and was a farmer. Mr. Alexander would have been 62 years old when he became a founding member. Clark was 93 years old when he died in October, 1999. Unfortunately, I have no other information on Mr. Alexander.

KEITH DUFF – Lived one mile southwest of the Buckley Show Grounds on four road. Keith owned a successful real-estate office in Traverse City. He was very involved with land purchases and later the land sales for Grant Farms. A large beef and land operation that operated in the Buckley and surrounding area through the sixty's and seventy's. Keith was significant in helping with some of the leg work when they were getting the Club established as a non-profit organization. Keith could be seen driving his mules around the show grounds pulling a wagon loaded with spectators. Keith was one of three signers for the original (non-profit) Articles of Incorporation for the Buckley Club.

Keith Duff: Founding member shown here with his team of pony mules hauling young spectators around the show grounds. Keith loved raising and training his mules. Keith was one of the members that signed the original Articles of Incorporation for the Buckley Club.

BASIL MOORED – Lived about twelve miles southeast of Cadillac, Michigan on M-115. Basil was an engineer and he owned his own steam traction engine. A 1910 Case 15-45hp that his father bought brand new. Basil restored it in his own shop. He was very knowledgeable around steam and was ready to share that knowledge with anyone interested. Basil also was an excellent plumber. He donated most of the plumbing materials and the plumbing installation for our clubhouse. Basil always had an interesting story. He was a huge asset to the Club and never missed a Club meeting. One of the meetings fell on Basil's 80th birthday, knowing he would not be at his own birthday party. The family brought the cake and ice cream to the Club meeting for all to share. Basil passed away in October, 1995 at age 84.

Basil Moored: Founding member shown here firing the upright boiler at the Buckley Engine Show. Basil was a dedicated Club member and never missed a regular Club meeting. He was very knowledgeable around steam and was always ready to share that knowledge.

THE FIRST SHOW AT COPEMISH

In 1967 the first Old Engine Show was a two-day event held on August 19th and 20th (Saturday and Sunday) on the Joe Rebman farm in Copemish, Michigan. The entry fee was a $1.00 donation with children under 12 accompanied by an adult free. Some of the first show attractions were three steam traction engines: 1917 Port Huron 24-75hp, 1910 Case 15-45hp and a Case 75hp, over a dozen tractors including the 1913 Little Bull 5-12hp, 1928 Rumley Oil Pull 25-40hp, 1921 Titan 10-20hp, 1928 Baker 25-50hp, 1917 Huber cross mount 12-24hp, 1917 Heider 9-16hp, antique vehicles, walk-behind garden tractors, hit and miss engines, grain threshing with steam engine, hand powered baler, corn shelling and grinding. Blough's sawmill powered by two steam engines, wood buzzing, drag saw and baker fan.

Joe Rebman's 1917 Port Huron 24-75hp.

With the first show being a success, the group decided to start pushing forward to become an officially recognized non-profit organization. On July 8, 1968 Leonard Clous, William Zue and Keith Duff signed the first (non-profit) Articles of Incorporation. On July 25, 1968 the Northwest Michigan Engineer

Basil Moored 1910 Case 15-45hp.

Ervin and Dan Blough's Case 75hp

In 1968 the second show was also held in Copemish, Michigan, on August 17th and 18th, 1968. It was agreed upon to hold all the future shows on the third weekend in August every year. The second show was more of a success than the first one. Some additional attractions to the show this year were, a 1917 Port Huron 19-65hp steam traction engine and a 1919 Stearns model "Q" 15-35hp tractor that was built in Ludington, Michigan. However, the newly formed Club found an enjoyable problem. More people wanted to join. This would bring in more steam traction engines, tractors, equipment, antique vehicles and in turn would bring in more paying guests. This would require more food venders, a larger parking area and more room for the show itself. Members were also coming up with ideas for new and well in this case older and better things. Expansion would be needed in the near future but how and where.

As the story goes, Leonard and a couple of the club members met with the Northwestern Michigan Fair board to possibly come up with a solution to their expansion problems. The fair was being boxed in at their grounds in Traverse City and eventually would be looking for a different location in the near future also. The hope was to possibly find a location both clubs could purchase together somewhere around Chums Corners, a few miles south of Traverse City. This way both clubs could share the cost of the land, some of the buildings, roads, etc. without one club having to bare the entire expense on their own. For privacy, each club would have built their own private buildings. These buildings would have been for storage, meetings etc. It could have been a win, win for both clubs. The Northwestern Michigan Fair board said, "They didn't think the Buckley Old Engine Club would ever amount to much and they didn't want to get involved with them". Several years later, the Northwestern Michigan Fair board decided it was time to find a new location outside of Traverse City. They set up a meeting with members of the Buckley Old Engine Club to see if they would be interested in sharing a location with them. Possibly, south of Traverse City near Chums Corners. That is when Leonard told the Fair Board that the Buckley Old Engine Club had grown to nationwide recognition. We did not want to jeopardize that by being involved with the small time Northwestern Michigan Fair that would probably never amount to much.

In the spring of 1969 James Breithaupt Sr., a Buckley farmer happened to stop at the Buckley Bar in Buckley Michigan. He was talking with Ken Deno the owner, their conversation got around to the Old Engine Show. Jim had attended both of the Copemish shows and commented on how crowded it was getting. Jim thought it would be better if the show could be held closer to Buckley. Ken said, "You know I own approximately eighty acres just west of here on the corner of west 13 Road and North 2 ½ Road. I might consider selling seventy acres of it to the club." Jim later got with Ray Clous, telling him what he found out. Ray took it to the board. The club, with loans from some of the members purchased the land. In 1969, the third annual Buckley Old Engine show was held on the newly acquired property and is still held there today. The purchase took place after the 1969 fliers were printed and distributed. The fliers still had the location of the show advertised to be at the Copemish show grounds. Thrashers Club, Inc. became an official organization. The mission statement was "*To promote the interest of the members of the Association in developing and in perpetuating the science and understanding of steam operated thrashers and other farm machinery.*"

In February 1969, the name was changed to the Northwest Michigan Engine and Thrashers Club, Inc. Then in June 1970, once again the name was changed. This time to the Northwest Michigan Engine & Thresher Club Inc. Doing Business As ("DBA") The "Buckley Old Engine Show". It remains to this day, the only other changes after this were to the clubs official addresses and when we received our exempt status, under section 501(c)(3). To help bring in funds for the newly formed club, members were required to pay membership dues of $1.00 per year, plus the gate entry fee at the shows. It was also in the bylaws that to be a member of the club you must also own and demonstrate at the shows something that operated with steam. However, as the club grew this law was sort-of overlooked although it was still in the bylaws until 1993. It was then changed to a club member being required to demonstrate at the shows an antique tractor or piece of antique equipment. Again as the club grew even this law was sort-of overlooked. Now to be a club member, you must be sponsored by someone that has been a member for at least three years and in good standings, be approved by the club's membership committee and agree to work the required hours at the show. You must also pay the initiation fee.

Leonard Clous 1913 Little Bull 5-12hp.

THE FIRST SHOW AT BUCKLEY

Carl Kreiser"s 1917 Huber Light-4

In 1969 August 16th and 17th, the first show was held on the new Buckley property. The show had all the room in the world for parking, demonstrations and everything else. It was now time to plan the layout of the grounds for future buildings, roads and show functions. Fortunately, the Hannah St. Mary's Church had loaned their large tents to the club so meals and some activities, such as arts and crafts could be enjoyed out of the elements. The St. Mary's Church also put on the charcoal broiled chicken lunch. The first structure built was a pit toilet located across the drive to the south of where the clubhouse now stands. The entry fee was still $1.00 with children under 12 accompanied by an adult free. Some of the attractions included, four steam traction engines, 1917 Port Huron 24-75hp, 1917 Port Huron 19-65hp, 1910 Case 15-45hp and a Case 75hp. Over two dozen tractors including the 1913 Little Bull 5-12hp, 1928 Rumley Oil Pull 25-40hp, 1921 Titan 10-20hp, 1928 Baker 25-50hp, 1917 cross mount Huber 14-24hp,

1919 model "Q" Stearns 15-35hp that was built in Ludington, Michigan, 1917 Heider 9-16hp, 1928 model "W" Rumley Oil Pull 20-30hp. Antique vehicles, walk-behind garden tractors, hit and miss engines, grain threshing with steam engine and horse power, corn shelling and grinding, horse powered baler, Blough's sawmill powered by two steam engines, Ray Clous' shingle mill, wood buzzing, drag saw and a baker fan. Irvin Luper and Tom Bartz set up a flea market along the driveway west of the east gate. Viola Kellogg was still there with her knickknack table. The flea market continued to grow throughout the years until it became one of the largest around.

Eli Truckner's 1917 Heider 9-16hp.

Leonard Clous 1928 Baker 25-50hp.

Joe Rebman's 1928 Rumley Oil Pull 25-40hp

NORTHWESTERN MICHIGAN ENGINE and THRESHER CLUB SHOW

August 21·22·23, 1970

1 mile west of Buckley

ALL RUNNING

- STEAM BICYCLE
- STEAM ENGINES
- GAS AND KEROSENE TRACTORS
- CROSS MOUNT HUBER, BULL TRACTOR
- RUMLEY OIL PULL, STEARNS TRACTOR MADE IN LUDINGTON, BAKER TRACTOR AND OTHERS. LARGE AND SMALL GAS ENGINES
- SAW MILL POWERED BY THREE STEAM ENGINES
- GRAIN THRESHING BY STEAM
- BAKER FAN, SHINGLE MILL
- CORN SHELLING AND GRINDING
- WOOD BUZZING, DRAG SAW
- HORSE POWER BALER
- VENEER MILL
- ANTIQUE CARS AND MORE

ANYONE OVER 65 FREE ON FRIDAY, AUGUST 21
ADULTS $1.00 – *Children under 12 Free if accompanied by Adult*

SHOW CLOSES 6:30 P.M. SUNDAY . . . FOR FURTHER INFORMATION CONTACT LEONARD CLOUS, Buckley
JOE REBMAN, Copemish

Lunch on Grounds

Leonard Clous 1920 Bates Steel Mule half track tractor driven by Harry Wieble

Steam powered bicycle.

Buckley show grounds in 1970, the cement block building is the bathrooms before the Clubhouse was added.

THE SHOW AT BUCKLEY 1970'S

In 1970 we had the first three-day show held Friday, Saturday and Sunday, August 21st, 22nd and 23rd. The entry fee was still a $1.00 donation with children under 12 free if accompanied by an adult and anyone over 65 free on Friday. The first major structure built was the bathrooms and showers put up in the spring/summer prior to the show. It was 36 x 24 feet with masonry walls and a flat roof. This later became part of the basement for the clubhouse. Some of the additional attractions to the show this year were, the upright boiler installed on the grounds to run all the stationary steam engines and miscellaneous steam items. Also there was Leonard's 1920 Bates Steel Mule half-track tractor, a steam powered bicycle, Bloughs sawmill powered by 3, yes, 3 steam engines, a goat powered treadmill machine and the veneer mill — its first year on the grounds. The veneer mill did not have a roof over it, and was operated by Morris Hulett.

John Sandula's 1915 Advance Rumley 20hp. Al Rhode driving.

Earl Hoch 1/8th scale Case 65.

Leonard Clous's 1925 Rumley Oil Pull Model L 15-25hp.

In 1971 the show was held August 20th, 21st and 22nd. A concrete slab was poured and a roof was built over the veneer mill. The veneer mill has been in the same spot on the grounds since it was delivered because this is where the tractor moved it when it ran out of traction and could not pull it any farther. Leonard said, "looks like a good spot to me". Some of the additional attractions to the show this year were: An Advance-Rumley 20hp steam traction engine, a replica 1/8-scale Case 65 steam traction engine, Leonard's 1909 Franklin air cooled car, a 1925 model "L" Rumley Oil Pull 15-25hp, a grain reaper, horse powered drag saw and a Calliope. The Buckley Volunteer Fire Department provided lunch.

Larry Wickern and Tom Graham's Case 65 ½ scale built in 1972.

Case 65hp steam traction engine

In 1972 the show was held August 18th, 19th and 20th. Some of the additional attractions to the show this year were, a Case 65hp steam traction engine, a 1972 replica ½-scale Case 65hp steam traction engine, many antique tractors, antique vehicles, hit and miss engines and there were carnival rides for the children.

A.D. Graham's 1922 A.D. Baker 19-60hp.

Joe Smith and Tom Graham's 1925 Nichols and Shepard 25-50hp. Later, Joe sold his share to Glenn Seagraves.

In 1973 the show was held August 17th, 18th and 19th. This year the fliers changed the wording to anyone 65 and over "FREE" on Friday. Ticket sales at the entry gate were just over $4,000.00. There were now eleven steam traction engines at the show including a 1917 Port Huron 24-75hp, 1917 Port Huron 19-65hp, 1910 Case 15-45hp, 1922 A.D. Baker 19-60hp, Advance-Rumley 20hp, Case 65hp, Port Huron 19hp, Nichols and Shepard, a 1972 replica 1/2-scale Case 65hp, a replica 1/8-scale Case and a 1904 Gar Scott. Some of the other additional attractions to the show were, a 1925 Nichols and Shepard 25-50hp kerosene tractor and we not only had grain threshing by steam and horse but believe it or not, we now had grain threshing by oxen.

Sam Zue and Vern Reynolds, with Vern's 1800's Sawmill.

In 1974 the show was held August 16th, 17th and 18th. The entry fee went up to a $1.50 donation with children under 12 free if accompanied by an adult. Anyone 65 and over "FREE" on Friday. Ticket sales at the entry gate were over $6,800.00. Vern Reynolds, one of our members, decided to retire and donated his 1800's sawmill to the club as a permanent fixture to replace the Blough's portable sawmill. Vern operated his sawmill for many years behind his home on West 8 Road, about four miles southwest of the show grounds. Vern, Sam Zue and crew operated the sawmill its first year under the clear (smoky) blue skies. Some of the additional attractions to the show this year were: an ox team with a cart that gave the spectators rides. This year a swap meet during the show was advertised on the fliers.

The food pavilion was the first major structure to be built on the grounds.

Carl Kreiser's 1922 Minneapolis 17-30hp. Later purchased by the club.

In 1975 the show was held August 15th, 16th and 17th. Ticket sales at the entry gate were just over $7,000.00. The first major building on the club grounds was the pavilion. It was built by Ron Bancroft and crew with material from the Wickes Lumber Company of Grawn, Michigan. The pavilions main structure was approximately 64 x 32 feet, with a 64 x 16 feet lean-to to the south. This later became the "Sam Zue Building". A roof was built over the immediate sawmill area. Sam Zue with his crew built it low to the ground to look like the mills from back in the day. Some of the additional attractions to the show this year were, a 1922 Minneapolis 17-30hp tractor eventually owned by the Club and an oil well pumper. The fliers advertised a camping area, swap meet and flea market. Ken Hittle was now doing the charcoal broiled chicken and Keith Galvin was doing an ox roast as Club members. We were starting to get a few other food vendors scattered about the grounds.

Terry Freese 1916 Russell 20-60hp.

In 1976 the show was held August 20th, 21st and 22nd. Ticket sales at the entry gate were over $7,400.00. There was a lean-to built for the horses to protect them from the elements. A roof built at the east end of the pavilion to cover the charcoal broiled chicken area (for you newer members, Oh! that is what that was used for). There was 16 feet added to the pavilion to the east bringing it right up to the charcoal broiled chicken area. Some of the additional attractions this year were, two additions to the steam traction engines. A 1915 Advance-Rumley 20hp and a 1916 Russel 20-60hp. A Monarch tractor, a steam operated car by the White Motor Company, a hot air engine, Hot Head engine, spinning wheel, miniature sawmill, well drilling, horse shoeing, grain cutting with a horse drawn binder and old time music. Another addition was A.D. Graham's 1/2-scale Oil Pull. A. D. also built a 1/2-scale Nichols and Shepard, 1/2-scale Titan 10-20 and a ½-scale Waterloo Boy.

Gene Cox on the hammered dulcimer.

A.D. Graham's ½ scale Nichols and Shepard. Now owned and driven by Carl Zue.

In 1977 the club's Tenth Anniversary show was held August 19th, 20th and 21st. Ticket sales at the entry gate were just over $9,700.00. A company called Lindenmuth and Associates Inc. wanted to lease ten acres off the southwest corner of the property to put in a natural gas sub-station and the club jumped at the chance. It was agreed there was no way the show would ever get so big as to need more than the sixty acres remaining. We could sure use the money to help with all the bills and for the funds needed to put on the next show. Within a very few years, we saw the error of our ways. With the club expanding by leaps and bounds, we wished we had never leased it. Some of the additional attractions this year were, a 1928 Greyhound 25-52hp tractor, a fiddler's contest and the mounted sheriff's patrol to help with parking.

Carl Kreiser's 1928 Greyhound 25-52hp. Later purchased by Fred and Ron Decker.

In 1978 the show was held August 18th, 19th and 20th. Ticket sales at the entry gate were over $11,000.00. The steam-powered bike did not return. When the weather broke in the spring, we invited Marion Kellogg to excavate the ground for the new clubhouse. Marion owned Kellogg & Son excavating. He also owned the property the club purchased in 2015 where we now do the plowing during the show. Marion excavated for the larger 36 x 40 foot open basement area that was connected to the west end of the bathrooms and shower area. The 36 x 64 foot clubhouse that Sam Zue laid out the floor plans for was then erected above them. There was a good-sized crew for this project because after they framed and stood up the east wall. They framed and stood up the entire 64 feet south wall in one piece. The club offered Ralph Pahl a farmer/roofer who lived on Zue Road a life membership if he would roof the clubhouse. Fritz, Leonard and Ray Clous installed the beautiful tongue and groove pine on the inside of the clubhouse. Ken Hittle (Leonard's son-in-law) trimmed the inside. Basil Moored installed the plumbing, Ken Morin installed the heating and mechanicals, Leonard installed the electric wiring and John Ockert did the drywall ceiling. We finally had a comfortable place to hold our meetings no matter what the weather. During the winter, the club bought a lowboy trailer with a hydraulic ramp and winch that was capable of hauling a big steamer plus an Oil Pull. It was an awesome trailer that would be perfect for moving the steamers. The club did not know the trailer had been previously stolen. When we registered for plates, its whereabouts were then known and the State Police confiscated it the next year. Additions to the show this year were, a 1928 Hart Parr 28-50hp tractor, many additional antique tractors, antique vehicles and over 150 small engines.

John Perry and his miniature sawmill.

Carl Peterson's 1928 Hart Parr 28-50hp.

Raising the trusses of the new clubhouse

The newly finished clubhouse, above the existing bathrooms.

In 1979 the show was held August 17th, 18th and 19th. Ticket sales at the entry gate were over $14,600.00. Steps were installed in the grade to the south of the clubhouse to make the passage a little safer. Some of the additional attractions this year were: a log cabin that was built in the 1930's south of Grayling, Michigan and brought to the grounds by Tom Small and others and placed under the large pine tree were it was used for the sound booth, a lean-to was erected east of the clubhouse for Ronald Blakslee (Pappy) our blacksmith and shingle brander, an eight-bottom prairie plow for the steam traction engines for large tractors to pull, popcorn made in a hog's kettle and the arts and crafts were allowed to move into the clubhouse basement.

Ron Blakslee "Pappy". The first blacksmith and shingle stamper.

Wayne Osborn with his Belgian horses and wagon full of spectators.

13th Annual Northwestern Michigan

ENGINE & THRESHER CLUB SHOW

AUG 15-16-17, 1980

ALWAYS THE THIRD WEEKEND IN AUGUST

BUCKLEY, MICH

MOUNTED PATROL ON GROUNDS

Old Time Music • Arts & Crafts Show • Large Flea Market

* Wooden Bowl Mill
* Veneer Mill
* Old Sawmill - Sawing most of the time
* Many kinds of large steam traction & stationary engines
* Nickols & Shepard, Rumley, Hart Parr, Baker and other large gas tractors
* Dozens of smaller farm tractors, including Heider, Bull, Bates Steel Mule, Others
* 100-200 small gas engines
* Antique 8 bottom prairie plow - plowing
* Spinning wheel in action.
* See grain cut in field by horse-drawn reaper, thresting by steam & horse power, ground into flour by gas engine and then baked into bread.
* Arts & craft show for your enjoyment and education.
* Baker fan - shingle mill
* Corn shelling & grinding with stone burr mill
* Horse powered baler.
* Antique Cars
* Miniature sawmill - working
* Working blacksmith
* Wood buzzing & drag saw
* Large flea market

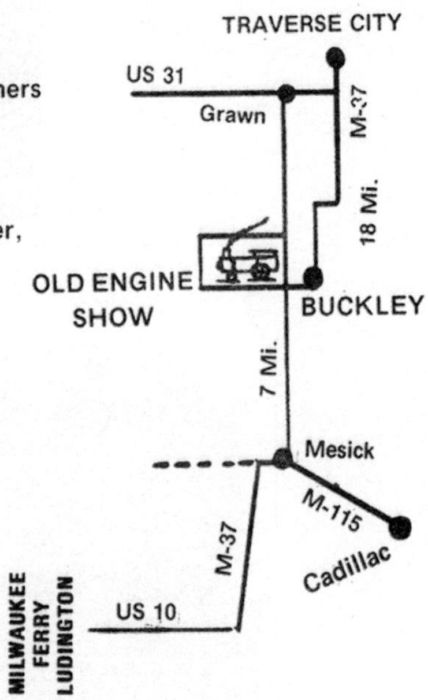

Parade Time -- 2:00 p.m.

Breakfast & Lunch on Grounds

Charcoal Broiled Chicken & Ox Roast

Anyone 65 and Over "Free" Friday, Aug. 15 only -- Show closes 5:00 p.m., Sunday. Adults $2.00 donation-Children Under 12 Free, if accompanied by Adult. For further information contact Leonard Clous, Buckley; Joe Rebman, Copemish; Tom Small, Roscommon.

In 1980 the show was held August 15th, 16th and 17th. The entry fee went up to a $2.00 donation with children under 12 free if accompanied by an adult. Anyone 65 and over "FREE" on Friday. Ticket sales at the entry gate was just over $18,000.00, (this was the first Saturday that sold more tickets than Sunday). This year the clubhouse had improvements that included an additional window installed to the east, two additional windows and an exterior door to the north with a raised wood deck installed all along the entire north side and a chimney for the wood stove installed in the open basement. Some of the additional attractions this year were: arts and crafts allowed in the upper level of the clubhouse plus the basement with an arts and craft show for your enjoyment and education, a windmill was erected to supply water for the steam engines, Keith Galvin's steam operated #9999 2-6-2 scale model locomotive, a treadmill run by a dog that churned butter and wheat ground into flour by a gas engine and then baked into bread. The fliers advertised there would be breakfast and lunch on the grounds.

The windmill that supplied water for the steamers.

Keith Galvan's #9999 2-6-2 scale steam locomotive.

In 1981 the show was held August 14th, 15th and 16th. Ticket sales at the entry gate were over $20,000.00. The food venders were the Bendon Eagles, Buckley Chamber of Commerce, Grant United Methodist Church, Buckley Senior Citizens and the Northwest Michigan Engine and Thresher Club. Some of the additional attractions this year were: the 55hp Anderson single cylinder Hot Head engine that was located, brought to the show and running the sawmill, all spearheaded by Sam Zue. There was an extension to the sawmill roof but did not cover the Anderson (the Anderson ran in the open its first year). The 250hp Corliss stationary steam engine that operated a flour mill in Grand Rapids, Michigan was roped off on the grounds under the clear blue sky its first year. The bowl mill for its first year was set up next to the upright boiler. A multi-station drinking fountain system with a trough below that flowed into a horse tank on the east side of the big hill powered by a hit and miss engine. Other attractions were a 1926 Avery 12hp tractor, a roof was put over the ox roast, and the 24" planer, the planer ran in the open its first year.

Bob Kellogg operating the Bowl Mill next to the upright boiler.

Sam Zue with the 55hp Anderson single cylinder Hot Head engine.

24" planer operated its first year without a building.

In 1982,the show was held August 20th, 21st and 22nd. Ticket sales at the entry gate was just under $22,400.00. This year there were seventeen steam traction engines at the show, a 1917 Port Huron 24-75hp, 1917 Port Huron 19-65hp, 1910 Case 15-45hp, 1922 A.D. Baker 19-60hp, a Case 65hp, 1972 replica 1/2-scale Case 65hp, a 2nd replica 1/2-scale Case 65hp, a replica 1/3-scale Case 65hp, a replica 1/8-scale Case, Nichols and Shepard, 1918 Minneapolis, 1904 Gar Scott, a 2nd Garr Scott, 1915 Advance Rumley, 1916 Russell 20-60hp, 1921 return flue Huber 22hp, and a 1931 Harrison Jumbo 20hp. Some of the additional attractions this year were: the Boy Scouts hot dog stand, the sawmill roof expanded to cover the entire mill area and the Anderson, the new steam building was erected around the newly acquired 250hp Corliss stationary steam engine, the planer building for the 24" planer (this building is now north of the woodworking building and used to house the track inspection cars, referred to as speeders), a 1937 Economy tractor (sold by Sears and Roebuck) eventually owned by the club. The fliers advertised primitive camping with no reservations. The club also held its first Swap Meet in the spring May 20th, 21st and 22nd.

Irvin Luper and Tom Bartz set up the first flea market along with Viola Kellogg, (not pictured).

In 1983 the show was held August 19th, 20th and 21st. Ticket sales at the entry gate was just over $24,000.00. This year the fliers stated "Dogs and bicycles allowed in the camping area only". The club purchased sixty acres of land across 2 ½ Road to the south (thirty acres from Art & Wilma Fall and thirty acres from Raymond & Rose Ann Quast) that helped immensely with our parking issues and traffic back-ups. Some of the additional attractions this year were, the bowl mill set up under the weather to the east of the steam building, the original horse lean-to had collapsed during the winter and was replaced with a new lean-to and some additional expansion was done to the sawmill roof. The previous year's Swap Meet was a great success so it was decided to continue with the spring Swap Meets.

Nick Lederle's 1904 Garr Scott.

In 1984 the show was held August 17th, 18th and 19th. Ticket sales at the entry gate was just under $29,800.00. We could now make apple cider with the purchase of a 1906 cider mill that came from Harbor Beach. To make things run a little smoother, the parade was held each day at 3:00 PM instead of 2:00 P.M. The show was expanding by leaps and bounds with more antique tractors, antique vehicles, small engines, arts and crafts, displays, food vendors and flee market vendors.

Mary's East Gate. Mary and Jim Arlt and crew.

In 1985 we had our first four-day show held August 15th, 16th, 17th and 18th. The gate entry fee was still a $2.00 donation with children under 12 free if accompanied by an adult. Seniors 65 years and older were free now only on Thursday and Friday. Ticket sales at the entry gate was over $34,100.00. The club purchased the 1918 Flour City 40/70hp prairie tractor. Iron -T-Junction was selling T-shirts and jackets on the grounds. St. Patrick's had their own food preparation building. A building was built for Mary's east gate and we added an 80 x 16 feet lean-to to the north side of the pavilion (that later became the "Sam Zue Building"). The club purchased the hog kettle we had been popping popcorn in.

Mike Clous manning the East Gate

Herb Kellogg at the camper shack, manning Herb's gate.

Geneva's gate. From left to right, Jerry Clous, Geneva Clous, Micky, Susan Kellogg, Catherine Weber, Chris Decker and Carol Clous.

This large oil well wheel capable of running 8 pump jacks was located outside its first year on the grounds.

Port-a-johns were placed on the grounds for the first time. By Sunday most of the members appeared to be pretty wore out as if they had worked an extra day at the show or something. In 1985, the Club was dealing with the city of Cadillac, Michigan to acquire the Shay locomotive which was on display in their city park. The engine did not fit in with the modern type renovation they were planning. At the last minute because of its historical value, the city, with the help of other groups, decided to have the locomotive restored and leave it right where it sat. The Shay locomotives were originally built in Cadillac Michigan but the high demand for the locomotive forced the factory to move to Lima, Ohio.

Sorghum Press located at the south end of the pond.

In 1986 the show was held August 14th, 15th, 16th and 17th. Ticket sales at the entry gate was over $56,500.00. The gate entry fee increased to $3.00 with children under 12 free if accompanied by an adult. Seniors 65 years and older were free only on Thursday. We had finally broken the $50,000.00 threshold at the gates. This year the campers brought in over $13,600.00 and the sound booth brought in over $6,500.00. In appreciation for her many years of service and for all she had done for the club, Geneva Clous (Leonard's first wife) received a well-deserved plaque. Her passing was a tremendous loss to the club. The clubhouse received a covered front porch at the west entry that included new steps and railings. The grade to the south of the clubhouse received new steps and a new handrail. Some of the additional attractions this year were: the 1918 Flour City 40/70hp prairie tractor was running for its first show, the Skinner steam engine was another added attraction and power supply for the sawmill system, the foundry was set up and operated in the steam building, a sorghum press was placed on the grounds south of the pond and a used boiler from Grand Rapids was purchased and installed in the steam building to assist the upright boiler. There were many additional antique tractors and antique vehicles, over 200 small engines and over 200 vendors in the flea market.

The club's 1918 Flour City 40-70hp

In 1987 the club's Twentieth Anniversary show was held August 13th, 14th, 15th and 16th. The monthly meetings were moved to the second Saturday of the month to help the members that had to travel long distances. The club purchased another 6.2 acres of land from Ken Dino (this is where the caretaker house would eventually be). This was one of the toughest shows we have ever had with rain almost constant for two of the four days and spotty for the other two days. The pond that was still in the middle of the grounds had overflown and you could have used a boat to row from tractor to tractor. The surprising thing was we almost took in the same amount of money as the previous year. Some of the additional attractions this year were, a 1930 Banting 20-35hp tractor and a large oil well pumping engine. In 1987, Elmer Pavlis owner of "This Ole' Farm Museum" on Pavlis Road, Buckley, Michigan wanted to work with the club to donate and move his museum to the club grounds with the option that he would be able to take guest through it during the show and by appointment in the off-season. A lot of time and effort by both parties went in to this adventure but in the end due to the outrageous amount of gift tax that would have been assessed on the donation it was not affordable and the project was abandon. Unfortunately, the family was forced to hold an estate auction for the museum's contents shortly after Elmer's death in 2005.

1930 Banting 20-35hp. owned by Glenn Arnold.

Rain can be a problem. Water, water everywhere!

In 1988 the show was held August 18th, 19th, 20th and 21st. The monthly meetings were moved back to the first Friday of the month. The club started planning for a better drainage system for the grounds. We dedicated the log cabin sound booth to Tom Small (the log cabin is now the "Small Chapel"). The much-needed oil well building was built by Grant Sheets, contractor. The "Buckley Cider Mill" building was built to house the 1906 cider press. The flea market received a new office building and the VFW building started out as just a small lean-to hut. Some of the additional attractions this year were, some very large engines that were used at oil well sites and facilities were located and running in the newly constructed oil well building. A drag saw cutting cedar logs into blocks for the shingle mill. The prony brake used for testing horse power operated under the clear blue sky off the northeast corner of the sawmill and the club owned 1917 Heider 9-16hp and 1929 Wallis 12-20hp tractor that we purchased from Eli Truckner's auction sale (Eli brought his Heider and other tractors to almost every show).

This log cabin was dedicated to Tom Small who brought it to the showgrounds from Grayling.

In 1989 the show was held August 17th, 18th, 19th and 20th. Some of the additional attractions this year were, Leonard's 1895 Peerless-Geiser 7 ½ hp portable steam engine (this engine once belonged to President Harding's father) that he purchased in 1987 from the Henry Ford Museum. Leonard used the Peerless to thresh grain with his wooden International Harvester hand fed threshing machine. Other attractions were: a 1929 Paige automobile, an Aultman Taylor 30-60hp prairie tractor and a 1924 Universal #3-rock crusher that came from Vanderbilt Michigan. We added free shuttle rides to and from the parking lot. We had over 400 small engines and the flea market had over 400 vendors.

Leonard Clous operating the 1924 Universal #3 rock crusher.

1895 Peerless-Geiser 7 ½ hp. Owned by Leonard Clous, and later owned by the Club

23rd Annual Northwest Michigan

Engine & Thresher Club Show

1990 4-Day Show

Aug. 16-17-18-19, 1990

Always the 3rd Weekend in August

BUCKLEY, MICH.

Old Time Music,(Non Electrified)
Arts & Crafts Show
Large Flea Market (Over 400 Dealers)

This Year Featuring The 2 Cylinder John Deere Tractors & Equipment

Anyone 65 and over "FREE"
Sunday, Aug. 19th

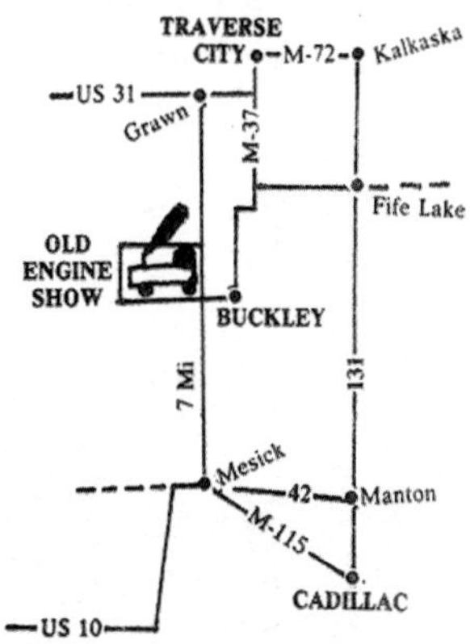

- Wooden Bowl Mill • Veneer Mill.
- Old Sawmill - Sawing most of the time
- Many kinds of large steam traction & stationary engines.
- Nickols & Shepard, Rumley, Hart Parr, Baker, Banting, Flour City, Aultman Taylor, Greyhound and other large gas tractors.
- Dozens of smaller farm tractors, including Bull, Bates Steel Mule, Huber, Heider, Wallis.
- 400+ small gas engines
- Antique Cars
- Antique 8 bottom prairie plow - plowing
- Prony Brake
- Wheat ground into flour by gas engine.
- Baker Fan - Shingle Mill - Sorghum Mill
- Corn shelling & grinding with stone burr mill
- Horse powered baler.
- Miniature sawmill - working.
- Working Blacksmiths.
- Wood buzzing & drag saw.
- Rock Crusher • Cider Mill
- No Alcoholic beverages In Show or Music Area
- Free Shuttles for Parking & Camping Area.
- Free Parking
- All Gates open at 7:00 a.m.
- All Gates close at 12:00 Midnight until 7:00 a.m. except East Gate for Emergencies Only!
- Non Denominational Church Services at 9:00 a.m. Sunday.

Threshing, Straw Baling Corn Shredding and Feed Grinding at 10:30 & 1:30 Daily

Parade Time 3:00 p.m. Each Day

No Alcoholic Beverages in Show or Music Area at any time.

– Generator Curfew at 10:00 p.m. –

- **Animals & Bicycles allowed in Parking Area ONLY (No Animals on Showgrounds)**
- **All Horses must be pre-authorized (Contact Wayne Osborne - Ph. (616) 378-2578)**

BREAKFAST & LUNCH ON GROUNDS -- CHARCOAL BROILED CHICKEN

Char-Broiled Hot Hamburgers -Soft Ice Cream - Pizza

Anyone 65 and over "Free" Sunday Aug.19th only. No Admission to grounds before 10:00 a.m., Aug. 15. Show closes 5:00 p.m., Sunday. Adults $3.00 donation; children under 12 free if accompanied by an adult. For further info. send self addressed stamped envelope to: **Ron Decker, 16585 230th St., Big Rapids, MI 49307 - Ph. (616) 796-2706; or Margaret Kreiser, 8488 Wilson Rd., Buckley, MI 49620 - Ph. (616) 263-7124 or Tom Graham, (616) 536-2700. Arts & Crafts-Delores Epley (517) 732-9400.** Show Phone: (616) 269-3669.

FOR FLEA MARKET INFO . . . Contact Rod Butler, 630 S. Frost Dr., Saginaw, MI 48603 - Ph. (517) 790-0846 Mary Arlt (616) 263-7441

In 1990 the show was held August 16th, 17th, 18th and 19th. The gate entry fee was still $3.00 with children under 12 free when accompanied by an adult. Seniors 65 years and older were now free on Sunday only. Membership dues went from $1.00 to $3.00, "outrageous". The club also for the first time in its history had a new president, Tom Graham. (Membership dues increase? New President? Naw must have just been a coincidence). Leonard became the club's first "President Emeritus". The music pavilion was built on the hill to the west of the sawmill where they could now also hold the Sunday morning non-denominational services. We erected the wooden windmill in the small engine area. A large tank was buried in the hill (to the west of where the woodworking building is now) to be used as an emergency water source for the fire department in case of a fire and also for watering the roads for dust control on the grounds during the shows. Some of the additional attractions this year were two Waterloo Boy tractors that were here for one year.

On the hill is the new music pavilion, to the left is the planer building and on the right is the VFW food building.

In 1991 the show was held August 15th, 16th, 17th and 18th. The gate entry fee was still $3.00 with children under 12 free when accompanied by an adult. We no longer offered a senior's free day. One of the largest undertakings of the club began with the group that started the process to bring the 1924 locomotive #207 to Buckley from Ypsilanti, Michigan, which meant eventually having our own railroad right here on the club grounds. A caboose, railroad ties and track were also purchased. There was thirteen acres of land purchased from Marion and Viola Kellogg on the north edge of the property to straighten out our north line for the railroad tracks. Sam Zue donated all the material to enclose the pavilion, because of this and for all he had done for the club, the now enclosed pavilion was named the "Sam Zue Building". The new power house building was complete (later dedicated to Wes Scott). The new foundry building was built to house the foundry and blacksmith. A house trailer was purchased and a well and septic installed for a caretaker who would monitor the grounds. The prony brake moved to the threshing area and included drawbar horsepower testing. A small hut was built for hospitality where they handed out programs and sold souvenirs, buttons and calendars. Additional attractions for this year were, a Stanley Steamer automobile, a steam tugboat engine, a flooring mill and the new whistle tower that keeps us on time. We had over 600 small engines. In an attempt to keep exhibitors on the grounds through Sunday afternoon, a $100.00 cash drawing was held at 4 PM on Sunday. The drawing had little success and after a few years was dropped.

Early version of the Hospitality Hut.

In 1992 the club's Twenty-Fifth anniversary show was held August 13th, 14th, 15th and 16th. The gate entry fee was still $3.00 but the wording was changed to children 11 and under free if accompanied by an adult. Elgin Strickland became the club's first grounds caretaker. A post-framed support and roof system was built over the caretaker's trailer by Sid Matthew's Builder. The music pavilion to the east was built to provide more room for the musicians. The full size railroad would be coming soon. Some of the additional attractions this year were: a 1914 Case 80hp steam traction engine, a 1918 Victor tractor kit (a kit that included a frame and wheels that you installed your own hit and miss engine on to create your own portable power unit). Iron-T-Junction set up shop in the steam building, we acquired a very large oil well pumping engine and Frank Hansen displayed his #79 all-wheel drive four-cylinder John Deere tractor that the John Deere Co. claimed they never built. The John Deere #79 was here as its last stop before going to the Smithsonian Institution where it would be on display for one year. After that, Frank and his wife toured the country with it going from event to event.

1914 Case 80hp. Owned by Larry Wickern.

In 1993 the show was held August 19th, 20th, 21st and 22nd. The pond that was in the middle of the show grounds had been removed and the drainage system was finally complete to hopefully alleviate some of the problems when we receive a significant rainfall (especially during the show). Some of the additional attractions this year were: the music pavilion to the west received a large addition to the north end, the large oil well engine turning the large wheel was set up in the oil well building (the large wheel was capable of pumping eight oil wells at one time with some of the wells being as far as a mile away). A roof was built over the prony brake giving the operators a safer work place out of the elements. Other additions were, a 1919 Frick 18-50hp steam traction engine and a 1908 Port Huron steam traction engine. A 1917 Aultman Taylor 30-60hp, a 1937 Ford air-cooled experimental tractor with an apposed 2-cylinder engine, a foot powered scroll saw, over 400 antique tractors and over 700 flea market vendors. The fliers specified no ATV's or golf carts, except for official use only. We also provided handicap transportation.

The large oil well wheel now locate inside the oil well building.

In 1994 the show was held August 18th, 19th, 20th and 21st. A new sound booth was erected west of the sawmill. The club purchased eighteen acres of land on the east side of 13 Road across from Mary's east gate from George and Mary Burnette. This area would be designated for members parking and a place to park the exhibitor's trailers. The railroad tracks were being laid by professional track layers. The wood working building site was being prepared. The bowl mill was moved to the west side of the oil well building. There was a new food building for the VFW and the electrical towers were constructed to supply better power throughout the grounds. Some of the additional attractions this year were: an 1898 Huber return flue 16hp steam traction engine and a 1924 Huber 25hp return flue steam traction engine. Leonard's 1909 Reo with an opposed 2-cylinder under the floor engine, a 1960 Hartz air-cooled tractor made in Germany, and a horse drawn air powered Calliope.

Horse drawn Calliope owned by Gerald Travis.

In 1995 the show was held August 17th, 18th, 19th and 20th. The railroad tracks were complete but unfortunately the train although working was not working well enough for the show. Some of the additional attractions this year were Leonard's 1886 Westinghouse portable steam engine purchased in Alberta, Canada, a 1911 Case 15-45hp steam traction engine. The wood working building was now in operation. The train depot was under construction during the show. The dairy booth moved from the flea market into the old St. Patrick's food building (the dairy booth also housed the first aid station). St Patrick's moved into their new building. The foundry added a lean-to to the south for the shingle stampers and the sawmill now had a roof and seating for the spectators.

A look inside the woodworking building

1886 Westinghouse owned by Leonard Clous. Later owned by the club

In 1996 the show was held August 15th, 16th, 17th and 18th. The entry fee was raised to $5.00 with children 15 and under free only if accompanied by a parent or guardian. There was also some strong discussions about going to a five-day show starting this year before the idea was voted down. Some of the additional attractions this year were: the train was running with lines of spectators waiting to ride, the train averaged over eight thousand riders per show. The train is by far our largest attraction. The kid's tractor pull was started and what a success it has been. The Buckley school offered showers for a small fee for anyone interested. The funds would go to help benefit their sports boosters.

The 1923 number 207 American locomotive first appeared in 1996.

In 1997 the club's thirtieth anniversary show was held August 14th, 15th, 16th and 17th. Because of traffic tie ups, the entry to the show was rerouted down County Line Road for those coming from the north and 4 Road for those coming from the south. Traffic then met on 11 Road and headed east on two and a-half road that was changed to one way just during the show days. This is how the traffic still enters today. Some of the additional attractions this year were, the 1911 Avery 40-120hp steam traction engine owned by Tom Graham, a 1919 Case 65hp steam traction engine and a 1904 Huber return flue portable steam engine. Power-take-off horsepower testing was added to the prony brake.

1904 Huber return flue, stationary engine. Owned by Wilbur and Charlie Volkening.

In 1998 the show was held August 13th, 14th, 15th and 16th. The log cabin that had been the sound booth in the past was now part of a club memorial honoring our deceased members and was dedicated to the memory of Tom Small as the "Small Chapel." Some of the additional attractions this year were, a new tractor shack located next to the flea market area. The first tractor slow race. Tractors of all types try not to cross the finish line first. The thresher crew put on a complete threshing demonstration in front of the big hill demonstrating the fine art of cutting grain with a scythe and cradle, to bailing the straw after they threshed it.

Leonard Clous's hand feed International Harvester Sterling threshing machine.

In 1999 the show was held August 19th, 20th, 21st and 22nd. The membership dues increased to $5.00. The "Buckley Train Depot" was dedicated to Howard Wilson, our first conductor. Another large undertaking for the club was moving the donated Mayfield Town Hall from its original location seven miles northwest of the show grounds and placed on its new foundation on the club grounds. It is used for a museum. The new machinery museum, dedicated to Leonard Clous, was built next to the music pavilion west of the sawmill. The new horse barn, dedicated to Wayne Osborn, replaced the horse lean-to. Our very first and hopefully last wedding was held on the grounds for Rick and Susan Fruin. Congratulations. The wedding took place in the coach car of the train. Afterward, the bride and groom were shuttled around the grounds and later road through the parade. Some of the additional attractions this year were, a 1913 Holt tractor that was here for one year and Dennis Kuhn's 1919 Napoleon truck that was built in Traverse City, Michigan.

The new Railroad Depot.

1919 Napoleon truck owned by Dennis Kuhn.

The Wayne Osborn horse barn and petting zoo.

★ 33rd ANNUAL ★

BUCKLEY OLD ENGINE SHOW

Northwest Michigan Engine & Thresher Club

August 17, 18, 19, 20, 2000

Always the 3rd weekend in August

BUCKLEY, MICHIGAN

Between Cadillac & Traverse City on M-37 Highway

Come see the PAST IN MOTION
"Specializing in Antique Steel Wheeled Tractors"

Fun for the Whole Family

Exhibitors welcome with Free Admission and RV Parking After 8 am Aug. 15th

Case Steam Engine

Large Flea Market
(550 Lots)
Flea Market Hours 9 am - 7 pm
Flea Market spaces by Pre-Paid Registration ONLY

OldTime Music
Arts & Crafts Show
Woodworking & Woodcrafts

Historic Harvesting Exhibit 1 pm Each Day on Parade Route

Kids Tractor Pull Ages 4-7 12:30 pm

No Animals Allowed in Show Area

Adults $5.00 Donation
Children *15 and under free ONLY*
if accompanied by a parent or guardian.

All Gates open at 7 a.m.
Gates close at Midnight
No New Admission after 9 pm
No New RV's after 7 pm–Holding area open until morning

Show closes after parade Sunday.
Free Parking & Free Shuttles to and from Parking Areas.
Handicapped shuttles until dark.

Tractor Slow Race Noon Each Day

Vendors & Non-exhibitor RV's welcome after 9 am August 16th

RULES & INFORMATION

- Sorry - NO ATV's or Golf Carts! Except for Official Use Only!!!
- Animals & bicycles allowed in parking lot & at your R.V. Site ONLY
- 24-Hour Security & Emergency Service. 911 Dispatch on Grounds
- Camp fires must be in approved container
- Generator curfew at 10:00 p.m.
- Showers at Buckley Schools to benefit Sports Boosters
- All Horses must be pre-authorized (Contact Wayne Osborne - (231) 378-2578)
- Curfew at 10:00 p.m. for 17 years old & younger
- No Alcoholic Beverages in Show or Music area at any time
- No Equipment Trailers in Campground
- Caretaker (231) 269-3678

FREE Rides on the Club's 1924 #207 Steam Locomotive

Threshing, Straw Baling, Horse Powered Baler 10:30 Daily

- Rumely, Hart Parr, Banting, Flour City, Aultman & Taylor, Greyhound and other large gas tractors
- Wooden Bowl Mill
- Veneer Mill
- Flooring Mill
- 1800's Sawmill
- Spark Show
- Rock Crusher
- Antique Cars (No Custom or Street Rods)
- Popcorn made in a Hog Kettle
- Wheat ground into flour by gas engine
- Prony Brake & P.T.O. Testing Open to all Tractors
- Corn shelling & grinding with stone burr mill
- Steam Plowing
- Memorial Chapel
- Non-denominational Church Services at 9:00 am Sunday
- Hundreds of Antique Tractors–Like Bull, Bates, Economy & Heider
- Avery, Huber, Frick, Case, Minneapolis, Port Huron and many other large steam traction engines
- 600+ Antique Gas & Oil Engines
- Working Blacksmiths & Foundry
- Breakfast, Lunch & Dinner on Grounds
- 250 H.P. Corliss Stationary Steam Engine
- Shingle Mill & Planer Mill
- Harness Maker
- Peerless, Huber & Westinghouse Portable Steam Engines
- Steam & Hot Air Model Table in Steam Building
- Petting Zoo
- And much, much more

Parade Time 3:00 p.m. Each Day

Free Parking

100-yr-old Museum Building

Cider Mill Demo
Daily, 10-11 a.m. & 1-2 p.m.
Sunday 10-11 a.m.

Show Information	*Show Information*	*Show Information*	*Flea Market Info*	*Arts & Crafts Info*	*Food Vendors Info*
Ron Decker	**Carol Clous**	**Jim Luper**	**Daphne Strickland**	**Irene Taylor**	**Bev Shillington**
16585 230th Avenue	7428 W. County Line Road	5430 Voice Road	11721 N. 13 Road	226 W. Church	7576 US 31 South
Big Rapids, MI 49307	Buckley, MI 49620	Kingsley, MI 49649	Buckley, MI 49620	Reed City, MI 49677	Grawn, MI 49637
(231) 796-2706	(231) 269-3053	(231) 263-5096	(231) 269-3678	(231) 832-9518	(231) 276-6420

Mary's East Gate
Show Phone • (231) 269-3669
After August 9, 2000
★ ★

In 2000 the show was held August 17th, 18th, 19th and 20th. Building also started on the Old Time Front Street located across the road from the railroad. The first buildings were the Harness shop where all sorts of leather goods are made and the Print shop with its 100-year old printing press along with many other styles of printing presses that print the club brochures and other such items. The railroads new water tower was erected near the tracks. A new flee market office was built by the north drive entrance and a modular home was purchased for the caretaker's residence to replace the house trailer. Some of the additional attractions this year were, the sawmill area is now completely covered including a very nice bleacher area with a backstop for the spectators. The 4-sided planer was at the show for the first time set up just north of the foundry building. The 4-sided planer was powered by a stationary Case power plant. A petting zoo was added to the horse barn and a Belgium horse named "Buckley" was the first horse to be born on the show grounds.

The Harness Shop located on Front Street.

The water tower at the train stop.

In 2001 the show was held August 16th, 17th, 18th and 19th. This was the first year for Education Days, a 2-day mini show put on by the club in October (education days are now held in September in hopes of warmer weather) where area fourth graders are escorted from one demonstration to another and learn the operation process of each station. Over 800 children attend the event each year. Front Street was expanded with the addition of the Old Time Necessities building. They can see demonstrations such as a spinning wheel, quilting, knitting, rope making, chair caning, whittling and more. During the off-season the Necessity building becomes storage for lightweight machinery. Some of the additional attractions this year were, a 1923 Huber return flue steam roller owned by the club. A 1928 Cletrac 40hp, a 1929 4-Pull tractor and an 8-cylinder model L-8 Cooper Bessemer engine added to the power house building.

The 4-sided planer in its first year at the Buckley Engine Show.

Students heading to their next demonstration at Education Days. Over 800 children participate

The Old Time Necessities building at the end of Front Street.

1923 Huber return flue steam roller owned by the club.

1929 4-Pull tractor owned by the Schuberg's.

In 2002 the show was held August 15th, 16th, 17th and 18th. The garage/machine shop was added on to the caretaker's residence in hopes that the members would be able to perform minor repairs to the club's equipment year round. Some of the additional attractions this year were: A 1912 Frick 9x10 steam traction engine, a 1923 Frick Eclipse Twin steam traction engine and a 1918 Minneapolis steam traction engine, the 1917 Hart Parr "Old Reliable" 30-60hp purchased by the club (the 1917 Hart Parr was the first machine to be called a "Tractor"), the 1938 Worthington 500hp diesel generator from Eaton Rapids, a tall wheeled log hauler loaned to the club and the 4-sided planer was moved to the east side of the sawmill.

1912 Frick 9x10. Owned by Gil Schrock

1917 Hart Parr "Old Reliable". Owned by The Club

In 2003 the show was held August 14th, 15th, 16th and 17th. This year there were 20 steam traction engines at the show, 1910 Case 15-45hp, 1917 Port Huron 19-65hp, 1922 A.D. Baker 19-60hp, Nichols & Shepard, 1972 replica ½-scale Case 65hp, a 2nd replica 1/2-scale Case 65hp, 1904 Gaar Scott, 1916 Russell 20-60hp, 1921 Huber Return Flue 22hp, 1898 Huber Return Flue 16hp, 1923 Huber Return Flue steam roller, 1931 Harrison Jumbo 20hp, 1911 Avery 40-120hp, a replica 1/4-scale Case, a replica 1/3-scale Case, 1914 Case 80hp, 1919 Frick 18-50hp, 1919 Case 65hp, 1908 Port Huron and a 1918 Minneapolis.

Ernie Allen's 1910 Case 15-45hp Thrasherman Special.

In 2004 the show was held August 19th, 20th, 21st and 22nd. The tractor shack that was located next to the flea market area was now moved to the front street area and the threshers received a much needed storage facility for belts, twine and other miscellaneous items. The Kellogg's allowed the club to do the plowing demonstration on their property that was attached to the north end of the club property ending a big problem with confinement in the show area. Additional attractions were a 1910 Case 15-45hp Thrashermen Special steam traction engine, a 1936 McCormick Deering WK40 tractor, 1927 Twin City 17-28hp tractor and a 1935 Parrett model 6 tractor. The power house acquired a 5-cylinder, 400hp, model 33 Fairbanks Morse engine. There were over 700 antique tractors on display.

1935 Parrett tractor Model #6. Owned by Kathie Thelen.

In 2005 the show was held August 18th, 19th, 20th and 21st. The year, part of our world came to an end when in March 22, 2005, the "Sam Zue Building" burned to the ground after thirty years of service. Miraculously the new "Sam Zue Building" was built and ready for this year's show (you would have thought Sam was still with us as quick as the new building went up). A Cobbler shop was built on front street were you can observe how shoes are repaired. A lean-to was added on the south side of the oil well building to house more engines and displays. A concrete floor was poured in the machine shop by the caretaker's residence. Some of the additional attractions this year were a 1933 Tructor tractor, a 1936 Kaywood tractor and two oil well pump jacks powered by the big wheel operating in different directions from the oil well building giving the spectators a better idea of how the system worked.

Terry Lohman 1927 Twin City 17-28hp.

Tall wheeled log hauler also called "Big Wheels"

In 2006 the show was held August 17th, 18th, 19th and 20th. Some of the additional attractions this year were: a 1903 E.W. Blitz curve dash automobile and tractors that included a 1919 Lauson 15-30hp, a 1949 Centaur, a Lauson 20-40hp, a 1940 Mayrath Standard and a 1949 Long model "A".

1936 Kaywood tractor. Owned by Carl VanWylen.

Lee Dodak's 1919 Lauson 15-30hp.

In 2007 the club's 40th Anniversary show was held August 16th, 17th, 18th and 19th. A new building was built for Mary's east gate with the driveway on each side for better control of who is coming in and going out. A used boiler built in 1950 by Kewanee Boiler Corporation was purchased and installed to replace the existing boiler in the steam building. The new shingle/planer building (later dedicated to Ray & Irma Clous) providing a safer work place with an enclosed area for the planer and a solid roof over the shingle mills. This building also provides storage during the off-season for the lumber cut in the sawmill, until there is a use for it.

Education Day guides at the east gate.

In 2008 the show was held August 14th, 15th, 16th and 17th. The 1907, 1,100hp Snow Engine (Snow Steam Pump Works of Buffalo, New York) was donated to the club. A group of club members went to West Virginia to dismantled it and see to its shipping to the club grounds. A huge undertaking for the 450,000-pound engine. Some of the additional attractions this year were, a 1922 AB Farquhar model "K" 20hp steam traction engine and an 1884 Westinghouse 15hp upright steam traction engine. A lean-to was added to the south side of the "Buckley Cider Mill" building along with remodeling the now interior wall so the guest could observe the cider mill process through plexiglas windows.

Rick Fruin 1903 EW Blitz curve dash automobile.

1907 110hp Snow Engine being transported from West Virginia to Buckley.

In 2009 the show was held August 13th, 14th, 15th and 16th. The new hospitality house was built in the area were the guests enter the grounds by shuttle. As they get off the shuttles, they can easily get information, directions and purchase souvenirs. The parade announcers received a new elevated booth west of the shingle mill/planer building. The Snow Engine is now on the grounds all 450,000 pounds of it. Plans are already in the works for its new permanent location and the type of building needed to enclose it. Some of the additional attractions this year were, a Barber shop on front street where you can get your haircut during the show and the sawmill crew split a large pine log the old fashion way using sledge hammers and hard work.

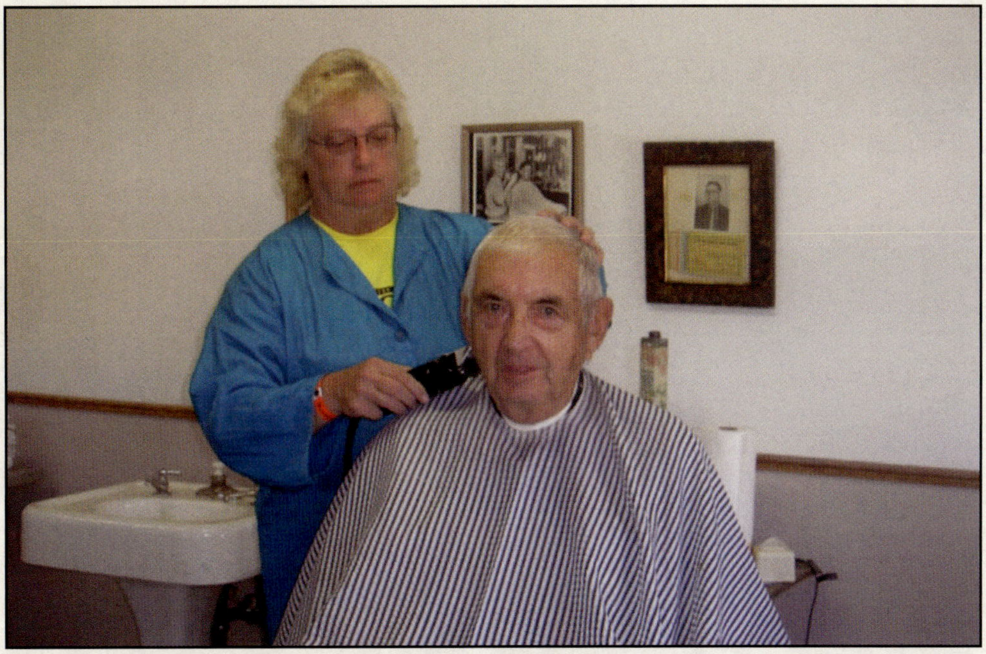

First person to get haircut, Herb Vanwylen in Tonsorial Parlor (Barber Shop) by Gloria Brufladt. The Front Street Barber Shop.

The new parade announcers booth.

The sawmill crew split a log the old fashion way with wedges and sledges.

In 2010 the show was held August 19th, 20th, 21st and 22nd. This year we purchased sixty acres of land from Ron and Ella Cochrane adjacent to the Club property to the west. One of the major projects this year was another building added to front street, the General store. You step back in time a hundred years and observe the thousands of antiques on display, along with penny candy, numerous snack food items and drinks for purchase. The bowl mill became a permanent fixture on the west end of the oil well building when a lean-to was constructed to keep the workers out of the elements.

The General Store located on Front Street.

In 2011 the show was held August 18th, 19th, 20th and 21st. A new shuttle building was built where the guest are dropped off on the show grounds. Two food pavilions were put up near the flea market as we rearranged some of the food vendors. Some of the additional attractions this year were the 1918 #7-steam locomotive restored by an awesome group of club members in five and a-half months having it ready just in time for the show. A 1914 Advance-Rumley 20hp steam traction engine, a 1926 Model-H 20-40hp Eagle tractor owned by the club, an 1870 Farar & Trefts Bessemer Conversion 10hp stationary engine, a Cooper-Bessemer model GAB 25hp stationary engine and three trailer loads (222) pedal tractors just an awesome display.

1918 #7 Steam Locomotive, an unbelievable effort by a special group of Club members that completely restored and had it running in less than 5 ½ months in time for the show.

1914 Advance Rumley 20hp owned by Wellington Farm Park

1926 Eagle model "H" 20-40hp owned by the Club. Driven by Brandon Jenkins.

In 2012 the show was held August 16th, 17th, 18th and 19th. This year we featured the National Ford/Fordson Collectors Association that brought over 200 additional tractors to our show. The association claimed this was one of the best shows they had ever attended. There was a major renovation to the drinking fountain area that took place this year with the addition of a working wooden water wheel modeled after the water wheel in Ravenna, Michigan and a wooden trough under the drinking fountains flowing into a wooden horse tank. Some of the additional attractions were, a 1937 Love tractor and a 1917 Fordson experimental 9X (the oldest Ford tractor).

222 pedal tractors, just an amazing display by Stan Emery and Larry Wiley.

In 2013 the show was held August 15th, 16th, 17th and 18th. The "Webber Barn" construction was started. Its original purpose was to house the Snow Engine but there was more in store for the club than just that. The 4-sided planer found its permanent location between the sawmill and the foundry when it was placed on a cement slab and a lean-to was built above it. Some of the additional attractions this year were, the purchase of the *Spirit of Traverse City* steam train from the city of Traverse City (a static display its first year). The train was becoming more of a liability than the city wanted to contend with. The train also did not fit in their plans for the open space renovation, part of which the train would have run through. Our feature tractor this year was the very rare Leader tractor and we had a 1948 Field Marshall tractor made in Great Britain that starts with the aid of a shotgun shell.

The new multi stage drinking fountain with wooden trough and water wheel.

Rich Buzalski's 1937 Love Tractor

The Spirit of Traverse City as a static display in the woodworking building its first year.

The Leader was our 2013 feature tractor

In 2014 the show was held August 14th, 15th, 16th and 17th. The "Webber Barn" was now complete and a dedication was held, there was a large turnout of club members on hand to show Mr. Webber and his Foundation their appreciation. It was also Mr. Webber's birthday so there was cake and beverages for all. The building also houses some farm chore, hands-on demonstrations. Everyone was able to see how they would be used during the show. The club installed a six-inch certified water well with pump house and water lines throughout the grounds along with a camper dump station as it forged forward to complete its licensed primitive campground for the working members. Some of the additional attractions this year were: the horse pulls sponsored by the Michigan Horse Pull Association. We held one of their best shows this year. The Snow Engine was progressing well with some of the heavy frame pieces set in place ready for grouting. The *Spirit of Traverse City* was running for its first show thanks to the long hours and back breaking work of our dedicated members who worked on the dirt moving, track, trestle, cars and engine.

New attraction for this year was the Horse Pulls.

In 2015 the show was held August 13th, 14th, 15th and 16th. The primitive campgrounds are now certified with a set of guidelines to keep the peace. The club also purchased twenty-six acres of land that Marion and Viola Kellogg owned, where the steam engines plow during the shows. There were four additional steam traction engines, a 1920 Advance-Rumley 18hp, a 1968 replica 1/4 scale Advance-Rumley, a 1914 Case 50hp and a 1912 Buffalo Pitts 15-45hp. Some of the other additional attractions this year were, the new depot for the Spirit of Traverse City. Lots of progress was being done on the Snow Engine. A 1911 International Harvester "Titan" 45 type "D" 27-45hp tractor and a 1916 International Harvester "Mogul" 8-16hp tractor. We held free speeder rides each day after the big train shut down (speeders are a name used for the inspection cars used by the railroads to inspect the rails). Some of the speeders were a 1937 Fairmont model 59C 8hp, 1943 Fairmont model M9 8hp, 1966 Fairmont model M19, 1925 Sheffield model 40-B 2-cylinder and a Buda open car.

The Webber Barn a tremendous donation by an awesome foundation, dedicated to Wayne and Joan Webber.

In 2016 the show was held August 16th, 17th, 18th and 19th. This year there were fifteen steam traction engines at the show, 1911 Case 45hp, 1914 Case 80hp, 1911 Avery 40-120hp, 1972 replica 1/2-scale Case 65hp, a replica scale Case, 1968 replica 1/4-scale Advance-Rumley, 1919 Case 65hp, 1923 Huber Return Flue steam roller, 1931 Harrison Jumbo 20hp, 1898 Huber Return Flue 16hp, 1914 Case 50, 1922 Farquhar model "K" 20hp, 1912 Buffalo Pitts 15-45hp, 1920 Keck-Gonnerman 18hp, 1907 Robert Bell. Some of the additional attractions this year were, the "Webber Barn" which got an interior face-lift with barn lumber along the south wall, a small mezzanine to give an old fashion barn scene and a hard wood floor in the upper office/storage room. The Snow Engine has all the heavy parts in the building and installed (thanks Team Elmer's for the lifts). Now it is time to increase the cleaning and scraping and preparations for painting, along with installing the hundreds of intricate parts that make up the rest of the engine.

1911 IHC Titan 45 type "D" 27-45hp owned by the Hudson Family.

1916 IHC Mogul 8-16hp owned by the Hudson Family

2017 will be the 50th year for the Buckley Old Engine Show and a lot of preparation is going into making this the best show ever, there is a good chance most of the steam traction engines and tractors that participated in the very first show in Copemish will be here for the 50th. The International Harvester Collectors of Michigan Chapter #11 will be the featured group and the Snow Engine is anticipated to be running via compressed air if all goes well. With the hard work from the dedicated members, it is progressing towards being the most spectacular show we have had yet.

1920 Advance Rumley 18hp owned by Marvin Blough.

1914 Case 50hp owned by Richard and Wes Doneth

1931 Harrison Jumbo Steam Engine on loan from the Fred Mida family.

1922 Rumely model G 20-40 on loan to the Club by Fred Mida family

THE RAILROAD

Buckley's Steam Locomotives and Rolling Stock

The first run of Buckley's 1923 American #207 4-0-4 in 1996 after a complete restoration.

1923 American steam locomotive #207 4-0-4

Buckley's first steam locomotive was used from 1996 to 2010 when it was replaced with a larger steam locomotive, the 1918 American #7 in 2011. #207 was moved to Nick Lederle's shop in Leland Michigan for restoration.

This is the unrestored 1923 American steam locomotive # 207 4-0-4.

#207 was moved to Nick Lederle's shop in Leland Michigan for restoration.

1918 American Steam Locomotive # 7

It was becoming evident that our 1923 American steam locomotive #207 4-0-4 was not big enough to handle the crowds at show time. We had traded the diesel locomotive that came with the Leelanau scenic railroad rolling stock for a 1918 American steam locomotive #7 that was dismantled but in good shape. In the spring of 2011, a dedicated group of club members started restoring that engine. In just 5 ½ months they restored Old Number 7.

This is the way the 1918 American locomotive boiler looked in the spring of 2011.

1918 American locomotive before restoration

1918 American locomotive under carriage

Fully restored 1918 American.

1918 American steam locomotive #7 in August 2011.

The Buckley Lantern 1998

'Reflecting on the Northwest Michigan Engine & Thresher Club'

The Buckley Railroad

By, Frank D. Holzschu

Slow, rhythmic blasts of steam. Gases drawn from the fire and pushed violently through the stack producing a puffy gray ribbon of smoke over the Buckley landscape. The Buckley passenger train begins its journey ascending the grade on the high iron rail of The Buckley Road. The No. 207 locomotive responds to the engineer's hand, easing the throttle as the train rounds the curve at the summit. The fireman works the water and steam valves initiating the injector, then hooks the fire with a soot blackened poker and shovels fresh coal over the flames giving life to the fire. The work caboose, supporting a large water tank on its platform, follows and functions as a tender for the locomotive. The train slithers past the Christmas tree farm and rounds the curve as the rail begins its decent. The passenger coach and covered gondola car are filled to capacity as the train glides past open farm fields on the left and a campground of tents and trailers on the right. The whistle screams a warning signaling the imminent road crossing ahead, the crossing guard smiles and waves as the train roars past. The conductor weaves his way between the seats in the coach thanking the passengers for riding The Buckley Road as the train rolls to a stop at the Buckley Depot.

The lantern's wick is turned up high with its bright light shining on the railroad, a railroad without a name. The railroad began in the early nineties when The Buckley Old Engine Club purchased a 0-4-0 steam locomotive. The members of the club had kindled a dream of building and operating a railroad on their showgrounds one mile west of Buckley, Michigan for a long time. Acquiring the locomotive was like turning on a draft inducer, drawing oxygen rich air through the flames of the dream. It took a short five years to build the road. The club's dream became reality on Wednesday, August 14, 1996 when engine No. 207 steamed its way over the road of rail, transporting a cargo of elated club members to the 29th annual Buckley Old Engine Show. The flames pushed the dream at such a rapid pace that little thought was given to a name. Two years later the railroad is still without a name. It is hard to write about something without a name and for that reason I'll give the railroad a handle, "The Buckley Road."

The Buckley Road began when Larry Wichern, Nick Lederle, Tom Graham and myself went to Ypsilanti, Michigan to prepare the 0-4-0 for the move to Buckley. The switch engine ended its working life many years previous, its major components were still intact but that's about all you could say for it. Time and the elements had worked their havoc. The smokebox lower portion was rusted through and the door was missing, the brake stand was missing, the gauges, the relief valve, the whistle and many other vital parts were missing. We worked one long hard day to get the locomotive ready to be lifted onto a lowboy tractor trailer. That evening we stayed at a nearby motel and we discussed the club's dream and talked about important railroad stuff. Larry spoke of the steam trains arriving and departing the depot in Cedar, Michigan when he was a boy, Nick told about the pros and cons of Sir Nigel Gresley's conjugating valve gear. Tom remembered going to the East Jordan & Southern roundhouse with his father to look at Engine No. 6 when he was young and I explained the desperate need for installing a track pan at Buckley so the train could take on water without having to stop. Now, most of these conversations had little or nothing to do with building The Buckley Road, but it's the very thing that caused the flames of the dream to dance and glow with a brighter hue. Not only were the four of us caught up in the dream, but almost every member of the club had their own story to tell that fanned the flames of the dream. Thinking back, a handful of traction engine plow boys, old engine and tractor enthusiasts thinking they could build a railroad, it kind of boggles the mind. It's a good thing we didn't know then what we know now or we may have left Ypsilanti without the locomotive. It takes a lot of time, it takes a lot of labor, it takes a lot of sweat, it takes a lot of phone calls, it takes a lot of patience and it takes a lot of money to build a railroad. We may have had some idea of this at the time but certainly not to the degree that we needed to know. The next day a crane lifted the locomotive onto the lowboy that carried it to Buckley and that was the beginning of The Buckley Road.

Nick Lederle (left) and Larry Wichern were involved in every aspect of building the railroad. They are two of the many people who turned the dream into reality. Photo by F.D. Holzschu.

It's been seven years since the flames of the dream engulfed the members of the club and many things have happened. Early on, a caboose was purchased and not long after that a Plymouth diesel switch engine came to Buckley. The club purchased rail, ties and spikes for the road and that's when the restoration of the locomotive began to get serious. The Alanson & Petoskey Railroad's rolling stock was purchased. The large diesel engine, its coach, work caboose and two gondola cars were moved to Buckley. A company was hired to lay the mile of track and a large pole building was erected to accommodate the train. In the first part of 1995 the track was finished and the locomotive was near completion. Anticipation was high and the thought of running the steam train at the 1995 August old engine show churned the flames of the dream among the membership. But it was not to be, little things, it seemed like an infinite number of petty problems began to plague the project. Club members tackled the problems one at a time with zest and vigor as the clock ticked toward the dead line and in the end, the clock savored victory. Over the next few months the problems were ironed out and the finishing touches were added. On Thursday August 15, 1996 the passenger train began service over The Buckley Road and the members of the Buckley club tasted victory. During the '96 and '97 shows the train transported over 20,000 people. A line formed at the Buckley depot each morning and persisted until nightfall. The faces in the line would change as the cars filled and departed for the twenty minute journey over The Buckley Road. The line shortened and lengthened and lingered and endured until the sun settled in the west, at day's end. A shadow from the train building would slowly swallow up the line, digest it for the night and then slowly regurgitate it as the sun rose in the east, the next morning. This happened each day of the two shows and there is no reason to believe it will be any different in 1998. Yes, the flames of the dream inspired many things to happen over seven years and the Buckley membership can stand up tall and be proud.

Many people worked hard to make The Buckley Road a success. Club members handled almost every aspect of the project. It is not possible to mention every person's name involved in the undertaking. Sheer numbers alone prohibit this and I simply do not know every person who had a hand in the dream. Still, some people need to be named and recognized. Ron Decker, Fred Mida, Tom Graham, Larry Wichern and Nick Lederle were key people. They were the people who kept the dream focused, they were the people who made the tough decisions, they were the people who stuck with the project through thick and thin and they were the people who spent endless hours behind the scenes making the dream reality. Also, in need of recognition are John Sprandel, Bob Jones, Wilber Volkening, Charlie Volkening and Glen Segraves. These men spent nights, weekends, holidays and for some taking time off from their business or job to make the dream come true. Likewise, in need of recognition are the anonymous club members that financed the project. Along with those mentioned above, there is one more group in need of recognition. A group of ladies indirectly helped make the dream possible, the wives of the men that gave so much of their time and talent to the project. Marge Wichern, Susie Lederle, Kathy Graham, Ruth Volkening, Helen Volkening, Davon Sprandel, Bonnie Jones and Mary Jane Segraves are to be commended for staying the course the many days and nights their husbands were off working on The Buckley Road. Marge Wichern speaks for the group as a whole, when she refers to the No. 207 locomotive as, "The other woman." A big thank you goes out to every club member who helped, supported and dreamed, for without you, The Buckley Road could not have become reality.

Few, if any, Buckley club members have set foot in a corporate board room or spent an afternoon at the country club. In most instances, club members have callused hands. He or she understands a hard day's work and is willing and happy to help a coworker, friend or stranger in need. Members have many talents and can be found filling a wide range of functions at the Buckley club, from the person busting coal for the locomotive fire to the president conducting a monthly meeting. It's not the job or status held that's important, it's the talents, the knowledge and skills possessed by the person that is important. A good coal buster never breaks more coal than needed, only enough to keep the immediate fire going. A good president knows when to talk and more importantly when to listen. The Buckley club has many fine coal busters and a good president. The work it took to build The Buckley Road did not happen by sitting around the country club lounge. The Buckley Road happened because people with dreams, imagination, determination, persistence and callused hands, worked together and stayed with the job until it was done.

The kerosene is running low and the lantern's light is starting to fade. It's time to drop "The Buckley Road" handle, and get back to a railroad without a name. I'm sure a name is coming. The Buckley Old Engine Club seems to handle the tough jobs in a short time when the draft inducer is turned on, drawing oxygen rich air through the flames giving life to the dream. The sun has set and the moon is rising over the Buckley showground's. Engine No. 207 leaves the depot again and ascends the grade of high iron rail. Slow rhythmic blasts of steam echo through the stillness of the night with the moon's soft light illuminating the puffy gray ribbon of smoke trailing over the Buckley landscape.

Many people worked on the restoration of Engine No. 207 turning a dream into reality. In the spring of 1995 a fire was placed in the 207 and she came to life for the first time in many years. The club members present for the occasion were: (Kneeling, left to right) John Sprandel, Tom Graham, Larry Wichern, Wilber Volkening and Frank D. Holzschu. (Standing) Frank L. Holzschu, Nick Lederle, Bob Jones and Charlie Volkening. (On the catwalk) Art Holzschu and Bryson Sprandel. Photo by Susie Lederle.

1926 Stillwell Passenger Coach

The Stillwell railroad coach before it was restored for use on the Leelanau Scenic Railroad. It was scheduled to be cut up for scrap but was saved when the man hired to cut it up ran into a nest of bees.

The Stillwell railroad coach before it was restored for use on the Leelanau Scenic Railroad. It was scheduled to be cut up for scrap but was saved when the man hired to cut it up ran into a nest of bees.

The 1926 Stillwell coach and the open gondola car.

The Stillwell as it looks now, giving rides at the Buckley Old Engine Show. Jim and Jan Dunmaier primarily restored the sheet metal sides on the coach.

Buckley's Wooden Caboose

Buckley's wooden caboose was made around 1910/1911. It was last used by the Central Vermont railroad. It was restored by Jim Ebel, Rick Vandenburg, Jim Siverly and Cindy Ebel. All the wood to restore the caboose was donated Dave and Mary Stark of Wexford Wood Works in Cadillac Michigan.

The club's wooden caboose before restoration.

Jim Ebel restored the caboose. Here he is pictured with his wife Cindy.

1912 Bucyrus Railroad Steam Crane, Ann Arbor #3187

The 1912 Bucyrus railroad steam crane served the Ann Arbor railroad for all most 70 years. It saw many derailments and train wrecks in its working life.

It was built by the Bucyrus Company in March of 1912. It has a lifting capacity of 100 tons. Its steam engine has two 10x12 inch cylinders that power its hoisting machinery.

The Ann Arbor railroad paid $13,675.00 in 1912 for the crane ($344,950.43 in today's money). The #3187 sits quietly on Buckley's railroad siding waiting her turn to be restored.

The 1912 Bucyrus Ann Arbor Railroad Steam Crane is a future project for the Buckley Old Engine Club.

THE BUCKLEY DEPOT

This first building built on Buckley's 1900's Front Street was the railroad depot. The depot was designed by Jim Saires and the frame was erected by White Oak Timber Company of Suttons Bay, Michigan. The timber frame was built during the 1994 show as an exhibit to demonstrate to the spectators how buildings were built 100 plus years ago. The frame was built using the mortise and tenon method.

There were no nails used in the frame construction. Jim Saires and his crew from the Buckley Woodshop finished the depot making it as historically accurate as possible. Now every passenger boarding the train gets a souvenir train ticket from the depots ticket window. The Buckley depot is dedicated to the memory our first conductor Howard Wilson and our first ticket agent Joe Kilpatrick.

Center part of the frame.

Final part of the frame.

THE SWITCH HOUSE

The Walton Junction railroad yard in early 1900's where the switch house sat.

This historic building was the railroad switch house in Walton Junction, Mich. Moved to the ground by Bob Voice and Dan Ockert and restored by Jim Saires and Gary Selander

RAILROAD SPEEDERS

In 2015 we added railroad speeder rides. After the 1918 steam train stops running for the evening, you can ride the rails on a speeder. Speeders are motorized railroad cars that were used by railroad workers to inspect and maintain the railroad tracks.

Steve Scott giving rides on a 1925 Sheffield speeder. The Sheffield is owned by Lee Lowing.

The club's Buda speeder

THE SPIRIT OF TRAVERSE CITY

The *Spirit of Traverse City* was the name chosen for this ¼ scale, 15 inch gauge, 4-4-2 Atlantic locomotive by its previous owner of 30 years, the city of Traverse City. The engine is currently owned by the Northwest Michigan Engine and Threshers Club and operated by its dedicated members. The history of this 68 year old engine, however, goes back much farther than its long time ownership by Traverse City and spans nearly 3000 miles across the country.

The beginnings to this 4-4-2 Atlantic, date back to the construction of its boiler, which was built by the Dan Russell Boiler Works of Boston, Massachusetts in 1948. The building of this scale, locomotive style boiler #111-79 was commissioned and purchased by Mr. Lawrence Witherill, also of Boston, Massachusetts. Lawrence, a railroad hobbyist, built the rest of the locomotive, tender and cars over the next several years. It was truly a remarkable achievement with which he took great pride in. Over the many years that followed, Lawrence would enjoy the fruits of his labor running the locomotive on weekends and holidays with his family at his home in Lexington, Massachusetts. Mr. Witherill's ownership of the engine would end with his death in April of 1966, after which the engine would go into storage under the care of a friend by the name of Walter Dunn of Lexington, Massachusetts.

After a short time in storage the little locomotive would be sold to Mr. Seymour Johnston of Santa Barbara, California. The engine was up and running shortly after making the 3000 mile journey to Mr. Johnston's estate. After only a small number of running days however, the cost of operation and idea of making a profit caused him to form a partnership with Mr. Gordon McLean of Tarzana, California. Once again the 4-4-2 Atlantic was on the move, this time to Whitter Narrows, California, where it was put into commercial operation, giving rides to the public. Unfortunately, due to a lack of security measures, the train was forced to be shut down due to vandalism issues after only about a year of running.

In 1970 Mr. Johnston sold the train to Mr. William L. Bay of Portola Valley, California. William held the locomotive in storage at his home for nearly a decade until he decided that he found an opportunity where the train could be enjoyed by everyone. Mr. Bay frequently visited his summer home in Rapid City, Michigan, during which he took interest in the city of Traverse City's future plans for Clinch Park, which at the time, had an operational scale railroad with a train powered by a gas driven locomotive.

William wanted to see a steam powered train running on the rails at Clinch Park; he also wanted the locomotive in his possession to be enjoyed by the general public. He presented his idea to the city board. Following some negotiations, the board voted to purchase the engine, tender, cars, and track from Mr. Bay. However, before the engine made the long journey to northern Michigan, Mr. Bay wanted to be certain that Traverse City would receive a great running train in tip-top condition. To make sure of this, the engine was taken to the Bay City Boiler Works of South San Francisco, California where the engine was completely disassembled and inspected under the close eye of the company president, Mr. John Greco. After thoroughly cleaning, sandblasting, and painting, the engine was reassembled with new components to replace any old ones that were not up to Mr. Greco's standards of quality.

The like new locomotive and train then made the nearly 2500 mile trip to Traverse City's Clinch Park in 1979. It became quite a task to get the train up and running at its new home, due to the higher than anticipated cost to install track and maintain the engine; the city board began to think it may have bit off more than it could chew. Complications resulting from attempts to have the locomotive's boiler certified by the state only made matters worse. Due to the fact that the boiler did not possess an A.S.M.E. stamp, the state of Michigan refused to certify the boiler because there was no proof that it was built to A.S.M.E. code. After obtaining a copy of the original boiler blueprints from the Dan Russell Boiler Works, the city was able to prove that although it lacked a physical stamp, it was still built to code and therefore was able to be inspected and certified by the state. Three years after the train first arrived in Clinch Park, the golden spike was finally driven by the city mayor, Mr. Larry Muzzarelli on July 19, 1982 and the locomotive was christened *The Spirit of Traverse City*.

The Spirit ran the rails at Clinch Park for nearly 30 years, becoming the city's emerald jewel steaming alongside the beautiful blue waters of West Grand Traverse Bay. It became a must see attraction to all visitors, as well as a weekend tradition to locals. Over the decades that followed, Traverse City took great care and pride in the little train that shared its name, until 2011 when the cost of maintenance and the Bay Front Planning Committee's visions for the future ended The Spirit's long reign at Clinch Park. The city announced, to the public's surprise, that it would not keep the train at the park because it did not go along with their plans for multi-generational park they envisioned. This was met with a great amount of criticism from the locals whom had grown fond of the train over the years. After weeks of discussion, the decision was made to sell the train; the locals emerald jewel was shut down for good. Bids were being sent to the city board from across the country, but no matter how far the engine traveled it was too far for the locals who loved it. With this in mind, the city made the decision to sell it to the Northwest Michigan Engine and Threshers Club, located just 20 miles south of Traverse City.

The Spirit of Traverse City was on the move for the last time arriving at the club's showgrounds in Buckley, Michigan in 2012. Before being put into operation, the train cars where rebuilt from the trucks up, air brakes were installed to help make the train safer during operation and a caboose was built to accommodate an air tank for the brakes as well as a seat for a conductor. The engine was given a new jacket and a fresh paint job as well as new fuel and water tanks. The fuel oil burners were redesigned and all of the old plumbing was replaced. Over the next two years, the grade was formed and the tracks installed, a trestle was built to overcome a dip in the landscape providing an aesthetically pleasing touch to the 2000 feet of track. The Spirit made its debut at the 2014 Buckley Old Engine Show, proving to be a big hit with hundreds of riders daily. However, during its first year of operation, it was discovered that the engine itself was in great need of repair. Over the following winter the engine was once again disassembled and completely overhauled at club member, Tom Graham's shop in East Jordan, Michigan. With the help of a handful of fellow club members, the restoration began. The cylinders were bored and fitted with new pistons and valves, and the entire valve gear assembly was given new pins and bushings. New axle shafts and redesigned bearings were fitted to the driving wheels that had been welded up and re-profiled. By the 2015 show the engine was running as good as ever. A station was added for storing the engine and to shade passengers where they patiently wait for a ride.

Once again at the show, thousands of passengers, young and old, ride along as *The Spirit of Traverse City* steams down the rails. The echo of the wheels clicking on the rails can always be heard as the train runs through the trestle, and the pull of the engine can always be felt as it chugs up the grade. Though the old 4-4-2 Atlantic has seen many changes, it is still giving riders just as much pleasure as it did when Lawrence Witherill first ran it decades ago.

400

FARM ANIMAL EXHIBIT

A trip to the Buckley Old Engine Show would not be complete without a stop at the animal barn. Pigs, chickens, sheep, rabbits, cows, and horses are just a few of the animal you can visit at the animal barn. One year we had a horse give birth to a beautiful colt at the show, and of course, they named him Buckley! Some years we have uncommon farm animals like llamas, alpacas, one year we even had an ostrich.

The piglets had a hard day at the Buckley Old Engine Show.

ANTIQUE CARS & TRUCKS

1942 Dodge Fire Truck donated to the club by Mike and Lilian Arrand and family of Cadillac, Michigan.

Leonard Clous driving his 1909 Reo

Tom Graham's Ford model T canopy Express.

Jan and Larry Clous with their 1921 Ruggles Model 20 truck on Front Street. This truck was owned and restored by founding member Leonard Clous.

1954 Chevy ½ ton pick-up truck restored by Bob Irrer. Owned by Barb & Ham Hobson

ARTS AND CRAFTS

We try to have something for everyone at Buckley. In the club house we feature arts and crafts. Here is a look at arts and crafts over the years.

One of the first crafters at Buckley was Georgina Martin. For many years she demonstrated her spinning wheel and loom.

Irene Taylor demonstrating the art of ceramics. Irene was also an arts and craft chairperson for the club.

MUSIC

Old time music has always been a part of the Buckley Old Engine Show. In the early days of the show, music was performed mostly in the evening along with square dancing. Today music and dancing take place most of the day in the big music pavilion and on the porch of the Sam Zue pavilion. We also have dulcimers playing in the evening in the Dulcimer Building over by the Oil Well building.

Gene Cox was the club's first music chairmen. Here he is in 1980 on guitar accompanied by his mother on dulcimer.

The Harmeling brothers preforming in 1981.

1906 CIDER MILL

A very popular exhibit at the show is the club's 1906 cider mill. The mill was made by the Hydraulic Press MFG Company of Mount Gilead, Ohio. The mill presses fresh apple cider at 10am and 1pm Thursday, Friday, Saturday and 10am on Sunday. You can get a free sample or you can buy a souvenir mug with free refills for life. The exhibit was originally operated by Bernie Depew & Leona Dowd with the help of Susan Decker. After Bernie and Leona passed away, Kenny & Sandy Hendges took over this historic and tasty exhibit.

Serving fresh tasty apple cider.

Bernie Depew at the controls of the 1906 cider mill.

POWER HOUSE

The Wes Scott Power House is the home of the 1938 Worthington six cylinder 500hp 300 Kilowatt engine and generator set. The Worthington came from the City of Eaton Rapids, Michigan where it and its sister engine were the first to be used in the Rural Electrical Association along with a hydroelectric plant on the Grand River. It was donated to the club by the Commonwealth Power Company.

Jeff Gunnett helping with the restoration of the Worthington engine. It looks like he is really getting into his work.

Steve Scott getting ready to start the 500hp 300 KW Worthington engine at the show.

55 HP ANDERSON HOT-HEAD ENGINE

We can power our sawmill a few different ways, with a steam or gas traction engine, the Skinner stationary steam engine, or our 55 horse power Anderson Type K Hot Head Oil engine. In May of 1982, founding member Sam Zue proposed that the club purchase a large one cylinder Anderson engine from Stan Forwood to add to our saw mill exhibit. The engine had powered a sawmill in White Cloud, Michigan. The sawmill burned down in 1950 and the engine had been sitting in the mill's back lot ever since. The engine was damaged but salvageable.

The club voted to buy the engine with club president, Leonard Clous saying, "we could have the engine moved to the show grounds for people to look at but there would not be time to get it running before the 1982 show." Someone sitting near Sam heard him say, "that's what you think?"

Sam and his crew went right to work. They poured a concrete pad at the saw mill to mount the engine. They poured many pounds of Babbitt for the rod bearings and did some machine work on the piston. A cooling tower was built and the air start system was repaired. Sam also designed a rather fancy exhaust system for the engine that blows perfect smoke rings when the engine is running. True to Sam's words, the engine was running for the 1982 show.

The Anderson Foundry and Machine Company of Anderson Indiana built their type K oil engines in one, two, three, four, and six cylinders. They were of semi-diesel design with a compression slightly lower than a diesel engine. To start the engine, a blow torch is used to heat a cast iron sphere at the top of the cylinder, hence the name "hot head." The single cylinder type K engine like Buckley's were built in 20, 25, 30, 45, and 55 horse power.

Skinner Steam Engine is one of the ways the club powers the saw mill. In the background you can see the Titysville boiler that provides the steam for the Skinner.

The Anderson Hot-Head

1927 GARDNER-DENVER STEAM AIR COMPRESSOR

In the steam building you will find our 1927 Gardner-Denver steam air compressor. The Gardner-Denver Company was founded in 1859 and is still in business today.

Buckley's Gardner-Denver compressor come out of the Acme Motor Truck factory in Cadillac Michigan. The Acme Motor Truck Company started as the Cadillac Motor Truck Company in 1915. In 1917 General Motors sued them for the Cadillac name. GM won that suit forcing them to change their name. The Acme Motor Truck Company went out of business in 1931. This historic steam engine was moved to the Buckley show grounds in 1977.

1927 Gardner-Denver steam air compressor.

1889 E.P. ALLIS CORLISS STEAM ENGINE

The club's Corliss steam engine was made by The E.P. Allis Company. E.P. Allis would become the farm machinery giant Allis-Chalmers in 1901. It is 250 horse power with a 16" bore and a 42" stroke. It runs at 80 RPMs. This engine powered the Voigt Flour Mill in Grand Rapids Michigan.

In 1865, Carl G.A. Voigt and W.G. Herpolsheimer opened a dry goods store in Michigan City Indiana. In 1870, they opened a branch house in Grand Rapids Michigan. In 1875 Mr. Voigt and Mr. Herpolsheimer bought a controlling interest in Star Flour mill in Grand Rapids. The mill flourished resulting in the partners purchasing the Crescent mills in 1882. The Crescent mill was located on the west end of Pearl Street on the Grand River in Grand Rapids. That is where this engine was used. In 1898 the Star Flour Mill and the Crescent Mills were merged to form The Voigt Milling Company. At the mills peak it produced 270,000 barrels of flour a year or about 800 barrels a day. The mill also dealt in corn, rye and oats.

Wes Scott saved the Corliss engine from the Voigt mill in the 1970's. The building was scheduled for demolition giving Wes only two weeks to remove the historic engine. While in the process of taking the engine out, Wes got sick with the chicken pox, slowing down the removal. He kept working to the point of almost being admitted to the hospital. The scrappers were standing by in case Wes did not make his deadline. Luckily he finished just as the demolition crews started on the other end of the mill. Wes crated the huge engine up and took it to his home in Greenville Michigan, where it was stored under an apple tree for many years. Wes always called the engine "the apple tree engine".

In 1981 Wes made arguments to bring the engine up to Buckley as a permanent exhibit. There were already two smaller steam engines, one from the Acme truck plant in Cadillac and one from the North Manitou saw mill and a Dallon boiler on exhibit. The Corliss would become the centerpiece of the new Steam Building. In the spring of 1982 concrete was poured for the foundation. That summer Wes and his son Steve started cleaning and sandblasting the engines parts. Steve remembers hauling the engine up to Buckley with his dad's old yellow one ton truck. He recalls that Cadillac hill on M-131 and Sherman hill on M-37 were almost being too much for the truck and the heavy load and they would have to pull over and let the rig cool down.

After the engine was installed and the new building finished, it was obvious that the old Dallon boiler could not provide enough steam to run the Corliss and other engines. So, a larger Kewanee boiler was added. Over the years more steam engines and the whistle tower have been added to the steam building, but the Corliss is still the crown jewel of the exhibit.

BUCKLEY'S PURPLE WHISTLE TOWER

This unique whistle tower was built on the showgrounds of the Buckley Old Engine Show. It stands 40 feet tall to the top of the jib crane and about 30 feet to the manifold on which the whistles are mounted. Wes Scott and Steve Scott conceived the idea for the tower in the summer of 1990 when a fellow whistle collector wanted to blow some of his smaller whistles at the show. His whistles were already mounted on a manifold, so we ran a line from the steam boiler out to his display and let some stream off. The display was a hit with the crowd. The next day Wes and Steve sat at the restaurant and started brainstorming on how to expand the exhibit and make it a permanent feature of the steam building.

The tower started out as a smokestack from a factory. Wes was going to use the stack as a culvert but abandoned the idea when the stack was crushed in the process of taking the smokestack down. Many days were spent straightening it and welding it back together. The top of the tower is a wheel, 6 feet in diameter which was formerly used on a conveyor line. It made a very functional standing platform. We then turned our efforts to the fun part, the manifold. We called all over inquiring if anyone could roll a piece of 4 inch schedule 80 pipe into a circle. Nobody we knew offered any workable suggestions so we took it upon ourselves to bend it into a circle. To our surprise after about an hour of hard work, we succeeded. Next we welded 14 pipe nipples ranging from 1 to 3 inches in diameter to the manifold to accommodate different size whistles.

Time was running short to have the tower ready for the 1991 show, so we did not have time to paint it. As it was, the crane was waiting to set up the tower when we arrived. Once erected, the tower needed only a steam line from the boiler to the tower itself, which turned out to be a good day's work. At last! The big test! Show time! We had 12 whistles on the tower the first year. The Wednesday before the show the boiler was hot and we had steam at the turn of a valve. At 5am Jim Gunnett and Steve Scott started blowing whistles (WOW did they get some attention!). People were coming from all over the showground's and from the restaurant a mile away to see what was happening. Our tower was a success. At the show everyone wanted to blow a whistle, so a schedule was decided upon. 6am wake-up whistle, 6:15am snooze whistle, 9am, 12 noon, 2pm and a 5pm whistle, also.

In 1992 founding club member Sam Zue made arangments to get the whistle tower painted. Without telling anyone he chose purple, not a color everyone was happy with but the public loved it.

DALLON STREAM BOILER

This vertical steam boiler supplied steam to the club's stationary steam engine exhibit for many years, until we need a bigger boiler. This boiler originally powered a laundromat in Cadillac, Michigan in the 1940s.

ERIE STATIONERY STEAM ENGINE

In 1908 the Smith and Hall Lumber Company of Traverse City Michigan began lumbering on the west side of North Manitou Island in Lake Michigan.

They built the village of Crescent with a saw mill, 600 foot dock and a rail line into the forest. By 1915 most of the land was logged and the operation was shut down.

We are very lucky to have one of the steam engines from that mill. It is an Erie stationery steam engine made by the Erie Engine Works of Pennsylvania.

GAS ENGINES

The gas engine area is dedicated to antique gas engines. Before electricity had reached rural America, the farm engine was a big part of farm life. It was a huge labor saving device. They were used to power buzz saws, water pumps, burr mills, corn grinders, washing machine and many other machine.

The manufacture of gas engines was a big business with hundreds of companies making many styles and house power sizes.

Gas engine exhibitors come in all ages. This young man is very proud of his Maytag engine.

Larry Mathews making corn meal with a burr mill.

Richard Nurmi next to his Superior Oil field engine

Bob and Barb Voice standing next to their IHC Famous gas engine

Old fashioned washing machine

Brownwall gas engine owned by Bob Hartwegs

BUCKLEY'S 1900'S FRONT STREET

Take a stroll down our 1900's Front Street and visit the blacksmith shop, cobbler shop, harness shop, Buckley print shop, barber shop and fountain. Visit the old time Necessities building and see spinning, weaving, rope making and try your hand on a loom and much more. Pickup a souvenir at the Hospitality shop. Stop in to the General Store and buy some penny candy or other old fashion goodies. Get your ticket at the depot to ride the 1918 steam train. See the Walton Juction switch house and the 1886 Mayfield town hall.

This clock used to be in downtown Elk Rapids, Michigan until a car hit it. Mike Winters asked if they would donate it to the club. Mike took it home, fixed the post, and recast the top to say our proper name. This really enhances Front Street.

THE PRINT SHOP & NEWSPAPER

The Buckley print shop.

The Buckley Print Shop started as a static display of a letterpress and a few other antique printing items owned by Jim Luper in the basement of the club house in 1988. The arts and crafts exhibit that occupies the upstairs of the club house needed more room, so the printing exhibit was moved to the oil well building. From there it was moved to the tractor shack. In 2000, the club built the harness shop building on the club's new 1900's Front Street and one side was the harness shop and one side would be the Buckley Print Shop. Dale Cobb and Jim Luper found more antique printing equipment to fill the new space. Jack Graham, Larry Johnson and Glenn Marshall joined the committee. In 2004, it was decided to build a new print shop designed by Jim Saires. This joined the harness shop, blacksmith and depot of Front Street. Later the Cobbler Shop moved into the old print shop space.

The print shop committee quickly added more old printing equipment and accepted donations to fill the new building. A letterpress, paper-cutter and supplies from the Lake City newspaper were added. A Heidelberg Windmill letterpress, Kluge letterpress and a Linotype machine were installed. In 2006, a 100-year-old Babcock newspaper press was donated by the Geiger print shop in Freeport, Michigan. This press was used to print the Freeport News and Clarksville Record newspapers. It is the print shop's intention to print a newspaper on this press someday.

As well as being a working exhibit at the show and at education days, the print shop prints a lot of material the club needs to put on the show.

The Buckley print shop crew Bridget Ennis, Jack Graham, Glen Marshall and Larry Johnson.

Jack Graham printing postcards on a Heidelburg Windmill letterpress

THE BUCKLEY BLACKSMITH SHOP

The Buckley Blacksmith Shop.

Ernie Edwards was also part of the Blacksmith team.

For many years we had a lady blacksmith, Darla Selander.

George Blair showing how the bellows work.

THE HARNESS SHOP

The harness shop, leather goods and the cobbler shop.

Rory Royston and Bill Anderson show how a harness shop would operate 100 plus years ago. A lot of the equipment was donated by Votruba Leather Goods of Traverse City, Michigan

Rory Royston and Bill Anderson showing how to make a horse harness.

THE GENERAL STORE

A look in the General Store.

THE NECESSITIES BUILDING

Kathy Bridson demonstrating the art of wagon wheel rugs

Arnold Wurm demonstrating a 100 year old rope maker.

Old fashion rug hooking by Jackie Putnam

WOODWORKING BUILDING

The Woodshop at Buckley started out in a corner of what is now the Oil Field Building, featuring mostly pedal and hand operated tools. Howard Bott, Carl Kreiser and Jim Saires used their contacts and networking skills to begin locating machinery.

In 1994 and 1995, a new building was built to house the woodshop and to house the train in the off season. The machinery is comprised of primarily flat belt pulley tools driven by our 100 feet of turn-of the-century line shaft technology (1800).

A collection donated by the Ameil Killman family consists of a 15-inch planer, 24-inch bandsaw and a belt drive mortiser. These tools are hand built from common angle iron, flat steel and some iron castings poured at Rounds Foundry in Traverse City in the 1930's. Buckley was lucky enough to also get the wooden patterns used for the cast iron parts which hang above the respective tools. To save money Mr. Killman built elaborate wood forms and cast the heavy base pieces in reinforced concrete which are painted machinery green for identification.

Much of the line shafting, pulleys, jack shafts and standards came from the Grand Rapids museum which came from furniture factories in the area that closed down in the 1970s and 1980s. The large stroke sander was donated from Manistee and the 24-inch Buss Jointer was donated by Brown Lumber in Traverse City. The club also owns a large lathe with wooden ways and legs, capable of turning porch posts up to 10 feet long, and 3 other antique unique lathes from the 1800 & 1900s.

This collection was gathered and installed primarily by Jim Saires, Carl Kreiser & Howard Bott. Many calculations were required to correctly install the right diameter pulleys to provide the proper machine RPM from the line shaft. Linkage to shift the belts from idler pulleys to the drive pulleys was engineered and built by three founders of Buckley's woodshop.

In addition to being a major exhibit at the Buckley Old Engine Show and Education Days, the woodshop has a dedicated crew of people and is used about 6 months out of the year for building and repairs needed on the show grounds.

Yates wood lathe being used to make a bowl by Carl Bott.

Oscillating belt sander made in 1895, operated by Ham Hobson.

This is the 1800's tredle powered jig saw restored by Jim Saires.

#3 Full Universal Wood Trimmer made by the Oliver Mchinery Co. There are ownly three of these known to exist. This one is on loan from Jeff Gunnett.

1926 Band Saw on the right of the photo, was made by Amiel Killman. One of four woodworking machine made by Amiel and donated by his sons Arlie & Wyn Killman. On the left of the photo is a 1895 friction drive drill press.

The 1916 Tannewitz Dowel Machine

On exhibit in the Woodworking Building is the 1916 Tannewitz Dowel Machine. It was made by the Tannewitz Works in Grand Rapids, Michigan. The dowel machine is owned by Jeff Gunnett.

This machine had been in a fire and was in rough condition when Jim Saires acquired it. Jim restored it to like-new condition.

The Tannewitz Dowel Machine takes square hardwood stock and makes it into round dowel stock. Each year at show time it produces around 3500 lineal feet of ¾ inch dowels and 800 eight-inch dowels for the kids during Education Days. You can see this amazing machine in action in the Woodworking Building.

Left to right Barb Hobson, Ham Hobson, Bill Sholten, Ron Schultz, Jeff Gunnett, Zach (Zippy) Parmenteer, Carl Bott

VENEER MILL

The club's veneer mill came from the Hiram F. Herrick basket factory in Ludington, Michigan where it was used from 1940 to 1965, to make veneer for fruit baskets.

Veneer refers to the thin sheets that are sliced or peeled off from hardwood logs. A veneer mill is essentially a large lathe. A log is held in place with a chuck or roller clamps at each end. The log is rotated against a carriage mounted knife on one side and a nose pressure bar on the opposite side to cut sheets of consistent thickness. The first few cuts that are obtained when the log is rotated produce sheets of varying lengths; this is called "round-up". This is usually discarded. Soon the rotating log starts producing perfect cylindrical sheets of wood akin to paper sheets from a roll. Besides being used to make baskets, veneer is most often used in furniture manufacturing.

This is a close up of how the log looks after the veneer has been peeled.

Here comes a sheet of veneer.

The veneer mill that came out of the Ludington basket factory.

THE SHINGLE MILL

Making cedar shingles with an 1881 Perkins shingle mill. The Shingle mill has a 38-inch blade that spins at 60 miles per hour.

Bob Kellogg and Larry Clous making wooden bowls for the show.

Steve Smith branding cedar shingles

YATES C-4A MOLDER 4 SIDED PLANER

Joshua Newlin and Tim Hartline with the Yates planer mill.

This planer mill was built in 1902 in Beloit, Wisconsin by the Yates-American company. The planer mill is a stationary machine that would be set-up in a building. In the early 1900s it may have even been powered by a water wheel.

In the years to follow, it made its way to a saw mill owned and operated by Jim and Marge Sanders in Stanwood, Michigan.

In 1977 it was purchased by the Hartline family. The Hartline's restored the mill to top shape and used it in their small wood working shop in Big Rapids, Michigan.

In was exhibited at the Buckley Old Engine Show in 2005 for the first time. It is exhibited at the four day show and at Education Days each year where it makes tongue and groove lumber. Much of the lumber planed by the mill is used for projects around the show grounds.

PLOWING

The club's 8 bottom P & O plow and Larry Wichern's 1914 Case 80 steam traction engine.

1913 John Deere 6 bottom owned By Frank D. Holzschu and Ken Billiau

Tom Graham's 1911 Avery 40 steam traction engine pulling his 1912 12 bottom P & 0 plow.

1912, 12 bottom P & O plow owned by Tom Graham

1918 FLOUR CITY PRAIRIE TRACTOR

There are only two Flour City 40-70 Special Road Tractors and eleven agriculture 40-70 tractors known to exist. What makes the agriculture model different from the Special Road model is that it has a mechanical power steering device, which enables the operator to steer the tractor under its own power, This made is much easier for road builders to use the tractor. The tractor has a horse power rating of 40-70. It has 40 H.P. on the draw bar (pulling) and 70 on the belt (running a piece of equipment with the belt pulley). Made by Kinnard-Haines Mfg Company of Minneapolis Minnesota.

4 cylinders, Stroke – 7.5 x 9 inches Traction wheels –8 foot x 24 inches
Front wheels – 48 inches x 10 inches Travel speed, miles per hour – 2 to 2.5
Fuel tank – 48 gallons Wheel base – 13 foot 6 inches
Width outside traction wheels – 111 inches
Fuel – Gasoline/Kerosene Shipping weight (approximate) - 21,000 LB.

1918 40-70hp Flour City at the Buckley show after its restoration and new paint.

In August 30th, 1985 the Northwest Michigan Engine and thresher Club purchased a 1918, 40-70hp Flour City prairie tractor. Leonard Clous, Raymond Clous, Carl Zue and Carl Kreiser with the permission of the Club and an approval to spend $20,000.00 headed to a Nebraska auction in Leonard's motor home. The auction was put on by the Hillpiper Auction Service out of Cedar Falls, Iowa. There was a group of over 100 riders that came on horseback and camped out as they participated in the auction. Fred Mida the Club treasurer was also at the auction; he had driven there by himself. The auction did not go quite as planned and the bidding on the Flour City was at $20,000.00 rather quickly and when the bidding was done the Club had purchased the Flour City for $26,000.00. There were a lot of irritated Club members when they learned how far over the set limit the purchase price was and a few members actually quit the Club because of what they claimed as irresponsible and foolish spending of so much of the Clubs money. When asked at the next Club meeting why they bid so much above the set limit Leonard the Club president and the designated bidder explained when the bid went over there limit of $20,000.00 he consulted with the other four members who were standing behind him on what to do next and as he turned back to the auctioneer someone behind him hit his arm swinging it in the air and seeing the movement and Leonard being an active bidder the auctioneer awarded him the next bid and Leonard claimed this action continued until they had the final bid. Leonard also claimed he had no idea who kept pushing his arm up to keep the highest bid and when he questioned the others no one was talking.

Club members at the auction purchasing the Flour City

1918 Rumely Oil Pull model F 18-35. Fred bought this tractor from the same auction that the club bought the Flour City tractor from.

During their trip the four members also attended the Old Engine Show in Mt. Pleasant Iowa taking in the sights and getting ideas for the Buckley show. Fred Mida for reasons unknown had driven himself to the auction turning down the invitation to ride with the others. Fred was there to purchase a model "E" 30-60hp Oil Pull and during that auction the bidding quickly went over Fred's limit and was purchased by another bidder for $35,000.00. Fred was so upset at the results of not getting the 30-60 that he headed for his car to leave having no other interest in the rest of the auction. Ray Clous one of the other Club members that had gone to the auction to help with the bidding on the Flour City had noticed a rare single cylinder

Oil Pull earlier and when he learned Fred was leaving he caught up with him and tried to encourage him to look at this other Oil Pull. It was a 15-30hp "F" and Ray tried to explain to Fred that it was a single cylinder Oil Pull and he had never seen one like it before and that it was rarer than the 30-60 Fred was originally interested in but Fred was still upset and adamant to be on his way when Ray jokingly said you know your wife is going to be really mad at you if you come home empty handed after all you went through to convince her to let you purchase the prairie tractor. Seeing that Ray was not going to give up Fred went with him to look at the "F" and the more he looked the more he agreed that this was the better tractor to own and when he finally admitted it Ray said you better do something quick because the auctioneer was about to accept the final bid so Fred jumped in and with his first bid ended up with the winning bid of $15,000.00. With the help of friends and neighbors Fred restored the 15-30hp Oil Pull and took it to the Buckley Old Engine Show and numerous times thanked Ray for encouraging him to buy it. Fred and his wife have since passed and with the generosity of their family members Fred's 15-30hp Oil Pull was once again at the Buckley grounds for the 49th Buckley Old Engine Show and negotiations are in the process to possibly keep it there for many years to come.

15-30hp Rumley Oil Pull, Model F. owned by Fred Mida.

<u>1886 Mayfield Township Hall finds a new home at Buckley</u>

Century State Building movers moving the 1886 Mayfield Township Hall.

For 109 years the Mayfield Township Hall stood at the corner of Center and Hannah roads in Grand Traverse County. In 1995 the historic building made a 6 mile trip to its new home at the Buckley Old Engine Show Grounds in Wexford County.

Built in 1886, it no longer suited the township's needs. It did not have restrooms and was not handicapped accessible. So, rather than tear down the old township hall they decided to build a modern Hall across the street from the old one. The Township donated the old Hall to the Buckley Old Engine club with the understanding that the club would pay for the move.

Central State Building Movers of South Boardman was contracted to move the 29 feet high 26 feet wide and 40 feet long 25-ton building. Club members lead by Buckley building committee chairman Ron Bancroft trimmed trees and held back branches while Consumers Energy worked to hold up power lines to allow the building to pass. At speeds of 5 to 15 mph it took 5 ½ hours to move the building the 6 miles to its new home. The Mayfield Township Hall now serves as a museum and the basement houses the Yesterday's home exhibit at the Buckley Old Engine Show.

FOUNDRY

The Buckley foundry makes cast aluminum dinner bells and Buckley commemorative plaques. The foundry is run by Mike Winters. Mike worked at the Traverse City Iron Works for 21 years and now owns Mike Winter's Casting in Elk Rapids Michigan. Running the Buckley Foundry is a family affair with his sons Myron and Larry, daughter Judy, son in-law Larry Masterson and wife Joyce all lending a hand. Club members Rich Numi and Brian Jefts have also help out.

They ram or pack sand into a mold called a flask to make the sand patterns in the shape of a dinner bell. Then aluminum is heated to 1300 to 1400 hundred degrees and the molten metal is poured into the mold. Out comes a ruff casting of a dinner bell. The bells are then finished and painted.

You can order a bell with your name casted in it to be delivered after the show.

The foundry has also cast parts for both of our steam trains and they have cast parts for many of our member's antique tractor, engine and car restorations projects.

Mike Winters forming the sand mold for a dinner bell.

Pouring the molten aluminum into the sand molted

SPRING & FALL SWAP MEETS

In the spring of 1982 the club added a spring swap meet to be held the 3rd weekend in May each year. Buckley's swap meet is a flea market for antique engine, tractor, tools and farm equipment collectors to buy, sell or trade hobby related items. It started as a three day event but is now a Friday and Saturday swap meet. It is a free event.

In 1990's A fall swap meet and antique car show was added to be held the 2nd weekend in September each year. The antique car show part of the event was not successful and only lasted two years. The swap meet part continued until 2015 when the club voted to discontinue it, but voted to bring it back in 2016.

Don Vyverberg was offering 5 hp. Fuller & Johnson engine for sale at the Swap Meet.

An unrestored engine looking for a new home.

Stover engine for sale or trade.

Fire Destroys the Sam Zue Pavilion

On March 22, 2005 fire destroyed the Sam Zue Pavilion. David Kreiser spotted smoke coming out of the air vents of the pavilion as he drove by the show grounds late Tuesday morning and called 911. By the time the fire department arrived, flames were shooting out of the roof, 30 feet in the air.

Built in the 1970's by Ron Bancroft, the pavilion was the club's main dining hall. Over the years, the building was added on to many times, expanding it to 4000 square feet with 3 commercially equipped kitchens. In the 1990s, Sam Zue, one of the founding members of the club, donated the money to add sides to the pavilion. This changed it from an open air building to enclosed one. To recognize Sam's generous donation, the club named the pavilion "The Sam Zue Pavilion."

The fire totally destroyed the pavilion and its contents. Stored in the pavilion for the winter was a 1922 Brockway Torpedo fire truck (the first motorized fire truck to be used on Mackinac Island), two stationary steam engines, four campers and a 1940 Farmall model H tractor. The fire truck was owned by founding club member, Carl Kreiser. It took Carl a few years but he was able to re-restore the historic fire truck. The Farmall tractor that was thought by many to be beyond restoration and was sold at our 2005 Fall Swap Meet. Remarkably the gentleman who purchased the Farmall tractor did restore it to like new condition and exhibited it at the 2006 Buckley Old Engine Show.

Through the hard work and determination of many people, a new Sam Zue Pavilion was constructed in time for the 2005 Buckley Old Engine Show in August.

The Sam Zue building on fire. The cause was never determined cause unknown.

1921 LaFrance Brockway Torpedo Firetruck pictured before the fire on page 11

The new Sam Zue building was ready for the August show 2005

BUCKLEY EDUCATION DAYS

In keeping with the Buckley Old Engine Clubs directive to preserve, show and educate the public about our agricultural and industrial past, Education Days was born in 2001. We invite local 4th grade students to spend the day at the show grounds, exploring and experiencing the exhibits first hand, learning more about how agriculture and industry has changed over the last 100 plus years.

In 2000 a committee was put together to organize the event and a call went out for members that would like to run exhibits. It was decided to make it a two day program. The first year we had a threshing exhibit, saw mill, prony brake, gas engine and drag saw, woodshop, steam building, old time necessities exhibit, harness shop, veneer mill, shingle mill and shingle branding, blacksmith, hands on farm exhibit and a steam train ride. The program has grown over the years to include a tractor exhibit, print shop, Sprit of Traverse City rides, rock crusher, barber shop, general store, Webber Barn, bowl mill, sticker mill, oil well exhibit and yesterday's home.

At the end of each day the kids, teachers, chaperones and bus driver are all treated with fresh apple cider and donuts. We send a goodie bag home with every kid and teacher, over the years the goodies have comprised of a show button, a dowel from the woodshop, a post card, bookmark and note pads from the print shop, a leather keychain from the harness shop, a branded shingle from the shingle mill, a bag of wheat from the threshing exhibit, a wooden bowl from the bowl mill, rope from the rope maker and train tickets from both trains.

Education Days is a free program to local schools making it very much in demand.

Students viewing the oil well exhibit from the cat walk.

John

Ham Hobson demonstrating the band saw in woodshop.

Lee Lowing showing how you hand start a John Deere.

Clara Porter serving popcorn to the kids as they walk by.

Washing clothes with a hand crank ringer washer.

Showing how to use an old fashioned coffee grinder.

A glimpse in the Webber Barn and some of the hands on farm exhibits.

KIDS OF BUCKLEY

Every day is full of lots of things for kids to do. Our guests can participate in a Scavenger Hunt from the Mayfield Town Hall that will take them around the ground to see what we have to offer. They learn about voting, by voting for their favorite exhibit. Then they can go downstairs to Yesterday's Home were they see what it was like to be a kid long ago. Next try their skills at a pedal tractor pull. Learn to make rope and weaving in the Old Time Necessities Building. Learn to drive a tractor at Tractor Driving School. At the Webber Building they can shell corn by hand and machine, grind corn, chop corn stocks, sort beans, learn how to weight produce and much more. Kids can ride on our train pulled by our 1918 full size steam locomotive and our ¼ scale steam train, The Spirit of Traverse City. Then spend some time in the sandbox, try their hand at old fashion games and play on a wooden train or model TT truck. Finish their day at the Animal barn.

Kids from 4 to 7 try to pull as far as they can.

Lots of encouragement from the crowd on a fine pull

After taking their turn pulling, each kid gets a prize.

Kirstin and Janelle Holzschu help get water for the train.

BUCKLEY TRACTOR DRIVING SCHOOL

In 2013 we started the Buckley Tractor Driving School. Spearheaded by John Crofoot and Norma Hilbrandt, this hands on exhibit gives people who have never driven a tractor a chance to do just that. After you learn how a tractor operates, you get to drive a 1939 John Deere Model H or a 1946 John Deere model B tractor. This exhibit is very popular with kids and adults. After your drive, you get a certificate to commemorate your accomplishment.

School's is in session. Kalobe Behrens learning the fine art of tractor driving.

EERE

Father and son, Mike and Shane Woodward learn to drive tractor at the 2016 show. They have the certificate to prove it.

ROCK CRUSHER

Cecil Butler bring a load of rock to the Universal # 3 rock crusher, with Larry Adams piching them in. The rock crusher was owned and restored Leonard Clous. He donated it to the club. We use the crushed rock around our grounds.

PRONY BRAKE

A Prony brake is a machine that is used to measure the horsepower of a steam traction engine or tractor. Horsepower is calculated by multiplying pounds (on scale) by RPM (on tach) and dividing by 1000. We can measure both belt horsepower and PTO horsepower on Buckley's brake.

An Amish family looks on as a tractor belt test is under way.

Farmall Tractor belt test.

THE WEBBER BARN

We would like to thank the Webber Foundation for this beautiful barn that houses our 1907 1100hp Snow engine and hands-on farm exhibit.

The Webber Barn was a new addtion to the show in 2015.

Mr. Webber shaking hands with Elmer Selander and club president Ron Decker.

Mr. Webber with the commemorative plaque for the new Webber Barn.

A Champion hair picking machine used to process horse hair for stuffing mattress on display in the Webber Barn.

JR Singer helping shell corn in the Webber Barn.

Webber Barn's old fashion kitchen.

Cream separator on display in the Webber Barn.

1907 1100HP SNOW ENGINE

The Buckley Old Engine Show's 1907 1100hp Snow engine was featured in the Gas Engine Magazine in 2016.

The club's 1907 1100hp twin tandem double-acting Snow engine is the only know example to exist of the historic engine, so of course, we had to save it. Bearing serial number C123 it was built in 1907 to pump natural gas from West Virginia to Pennsylvania and New York. Originally installed in a natural gas company pumping station in Hunter, West Virginia. It worked there until about 1944, when it was dismantled and moved some 90 miles south to Gulf pumping station in Ritchie County. In 1967 the station was retired, and this engine along with a 2nd identical engine number C124, was sold to local oil and gas producers H. H. Elder and Son. The new owners bought the engines for the building they were housed in to use as a storage facility for their business.

The Elders spent the summer busting up engine number C124 for scrap, but spared engine C123 as their Grandfather wanted to see the engine restored some day.

The engine sat undisturbed for the next 41 years. In 2007 the Elders started looking for a suitable home for engine number C123. There looked through the Farm Collector Show Directory for an engine club larger enough to accommodate the huge Snow engine. They wanted to donate it to a club that would get it running, that is when they found the Buckley Old Engine Club.

The club sent club members Tony Suykerbuyk and Steve Scott to West Virginia to look at the engine. Tony and Steve went to see the engine early in the morning, with the engine being so large and so far from Michigan they figured the club would not want it. They ended up staying the whole day and decided the engine needed to be saved.

In 2008 disassembly started and by mid-2009 the engine was safely on the Buckley showground's Besides the full size railroad this has been the largest project Buckley has ever undertaken. Now housed in a beautiful building donated by the Webber Foundation the Snow is undergoing a full restoration.

The 1907 1100 hp Snow engine

Steve Scott cutting the nuts and bolts so the flywheel can be removed.

The crew of members starting to dismantle Snow engine in West Virginia in 2008.

West Virginia crew

Lowering half of the 18 foot diameter flywheel into position.

Installing the cylinders.

FROM STEAM TO DIESEL

The largest exhibit area at Buckley is the antique tractor area. With over 1000 antique tractors in attendance, you can see how tractors have evolved over the last 120 years from steam to kerosene to gasoline to diesel. With makes that most people have heard of like John Deere, Farmall, Ford and Case to more obscure brands like Bull, Bates Steel Mule, Four Pull, Friday and Economy. You can see tractors in action too. Plowing, powering a threshing machine, testing their horse power on the Prony Brake, or on parade. We also have tractor games like the tractor slow race and the tractor poker run.

The Tractor Shack is were we register all the exhibitors. We also park all the tractors, run the tractor slow race and the tractor poker run.

The Bull Tractor Company of Minneapolis, Minnesota started selling tractors in 1913 and by 1915 was America's top selling tractor. Farmers wanted a smaller tractor, suited for smaller acreage farms. On paper the Little Bull fit the bill. In company advertising they claimed "The tractor sensation – does the work of five good horses and sells for the price of two poor ones" "It's light and can travel anywhere" "Very durable" "The Bull with the pull". But farmers that bought the Little Bull felt very different. With its 5-12 hp and one wheel drive, it was under-powered. It had an opposed two cylinder engine and open gears that were exposed to the dirt and dust of the field. It broke down in record numbers, and most were returned to the manufacturer

1907 Robert Bell steam traction engine, owned by Mike Besko.

This 1897 Huber steam traction engine (steam tractor) is the oldest tractor at the show. Owned by Charlie and Helen Voelkening. This engine was under water in the back waters of the tippy dam for 41 years.

Unloading a Keck-Gonnerman steam traction engine owned by Marvin and Liz Blough at the 2016 show.

1914 Advance steam traction engine, owned by the Wellington Farm Park of Graying.

1917 Hart-Parr 30-60 tractor, owned by the Buckley Old Engine Show.
Hart-Parr was the first tractor company to use the word tractor.

Ron and Chris Decker's 1917 Aultman & Taylor 30-60 arriving at the show.

This 1925 Nichols & Shepard 25-50 was a Buckley staple for many years. Even though it is a gas tractor you can see its steam engine roots. Owned by Tom Graham and Glen Seagraves.

Rumely Oil Pull model G 20-40, owned by Tom Hoose. The model G was made from 1918 to 1924. Rumely Oil Pulls are cooled with oil instead of water, hence the name Oil Pull. Rumely's moto was burns kerosene under any load.

1926 Rumely Oil Pull model M 20-35 owned by Larry & Norma Hilbrandt. The model M was built from 1924 to 1927 with 3671 made. In 1926 it sold for $2000.00.

John Deere's first general propose tractor the Model GP. Made from 1928 to 1935. This 1928 model is owned by Lee Lowing

1926 Fordson model F tractor owned by Ron and Nelda Fredrick. The Fordson was Henry Ford's first entry to the tractor market. Henry Ford used his famous assembly line to mass produce his tractor. From 1917 to 1928 there were 552,799 American Fordson tractor made. For many years the Fordson dominated the tractor market.

1926 Eagle model H 20-40. This tractor is owned by the Buckley Old Engine Show.

This 1929 Hart-Parr 28-50 is owned by Joel VanAntwerp. It has been in Joel's family since 1936. His great-grandfather bought it used to farm with and to power a saw mill. Joel's grandfather, Carl Peterson, started showing it at The Buckley Old Engine Show in the 1970s. Now Joel carries on that tradition.

The Four Pull tractor was made in Big Rapids Michigan from 1915 to 1930. The Four Pull was not the first four wheel drive tractor but it was the first to use gears instead of chains to transfer the power from the engine to the axles and wheels. There are only five Four Pull tractors known to exist. This one is owned by the Schuberg family.

The Farmall Regular made by International Harvester is credited with being the first successful row crop tractor. The Regular was built from 1924 to 1932. This 1928 model is owned by Greg Beerns.

1929 Rumely Oil Pull model Z 40-60 Super Power. The last tractors to carry the Oil Pull name were the Super Powers series. The Super Power series consist of the model X, Y and Z. The model X was made from 1928 to 1929 and the models Y and Z were made only in 1929. There were only 240 Model Z's made. This rare tractor is owned by Bob Crandel

1935 Allis-Chalmers model E 25-40. Produced from 1930 to 1936 with only 1425 made. On October 13th 1932 Allis-Chalmers was the first tractor maker to sell a tractor with rubber tires as standard equipment. This changed farming forever.

1930 Banting 20-40 owned by Glenn Arnold. Allis-Chalmers manufactured this tractor for the Banting Company.

1939 Economy tractor powered by a Ford model A car engine. The Economy tractor was sold through the Sears, Roebuck Department stores catalog. You could order it with an engine or with out and put your own Ford engine in it. The Buckley Old Engine Club bought this tractor from long time member Leonard Amidon.

Leroy Rohn on a 1939 Ford 9N. Leroy was always an enthusiastic exhibitor. His sons are carrying on the tradition

1941 McCormick Deering O-4 orchard tractor owned by Scott Fales

Norma Hilbrandt on her 1945 Farmall model A. The tractor has been in Norma's family since it was new. She is pulling a Hercules buzz rig in the daily parade.

1945 Cletrac model HG crawler tractor owned and restored by the Schudlich brothers of Manistee.

Many people make scale models of tractors and engines Cody Cook built this homemade version of the Avery when he was a teenager. He called it the Favery. (Cody Not pictured)

1966 Case model 1200 owned by Stanley Groner

1951 Sheppard SD3 diesel tractor made by the R.H. Sheppard Company of Hanover, Penn. The SD3 was the three cylinder model. Owned by Don and Cathy Nutt.

INDEX

1/2-scale Nichols and Shepard 31
1/2-scale Oil Pull 31
1/2-scale Titan 10-20 31
1/2-scale Waterloo Boy 31
1/8th-scale Case 65 25
15-30hp Rumley Oil Pull, Model F. 148
55hp Anderson single cylinder Hot Head engine 39
#79 all-wheel drive four-cylinder John Deere tractor 54
250hp Corliss stationary steam engine 39
1870 Farar & Trefts Bessemer Conversion 10hp stationary engine 78
1884 Westinghouse 15hp upright steam traction engine 74
1886 Westinghouse portable steam engine 57
1895 Peerless-Geiser 7 ½ hp 49, 50
1897 Huber steam traction engine 182
1898 Huber Return Flue 16hp 69, 85
1903 EW Blitz curve dash automobile 72, 74
1904 Gaar Scott 28, 40, 41, 69
1904 Huber return flue portable steam engine. 59
1907 1,100hp Snow Engine 74
1907 Robert Bell 85
1907 Robert Bell steam traction engine 182
1908 Port Huron steam traction engine 55, 69
1909 Franklin air cooled car 26
1909 Reo 56, 109
1910 Case 15-45hp 14, 18, 28, 40, 69
1910 Case 15-45hp Thrasherman Special 70, 69
1911 Avery 40-120hp 59, 69, 85
1911 Case 15-45hp steam traction engine 57
1911 Case 45hp 85
1911 International Harvester "Titan" 45 type "D" 27-45hp tractor 84
1912 Bucyrus railroad steam crane 99
1912 Buffalo Pitts 15-45hp 84, 85
1912 Frick 9x10 steam traction engine 68
1913 Holt tractor 61
1913 Little Bull 5-12hp 5, 14, 18
1914 Advance Rumley 20hp 78, 79, 183
1914 Case 50hp 84, 85, 86
1914 Case 80hp 54, 69, 85, 144
1915 Advance Rumley 20hp 24, 31, 40
1915 Bates Steel Mule half-track tractor 24
1916 International Harvester "Mogul" 8-16hp tractor. 84
1916 Russell 20-60hp 31, 40, 69
1917 Aultman Taylor 30-60hp 55, 184
1917 Cross Mount Huber 14-24hp 18
1917 Fordson experimental 9X 80
1917 Hart Parr "Old Reliable" 68, 184
1917 Heider 9-16hp 14, 18, 19, 48
1917 Huber cross mount 12-24hp 14

1917 Huber Light-4 18
1917 Port Huron 19-65hp 17, 18, 28, 40, 69
1917 Port Huron 24-75hp 9, 14, 18, 28, 40
1918 American steam locomotive #7 90, 91
1918 Flour City 40/70hp prairie tractor 43, 45
1918 Minneapolis 40, 68, 69
1918 Rumely Oil Pull model F 18-35 147
1918 Victor tractor kit 54
1919 Case 65hp 59, 69, 85
1919 Frick 18-50hp 55, 69
1919 Lauson 15-30hp 72
1919 Stearns model "Q" 15-35hp 17, 18
1919 Napoleon truck 61
1920 Advance Rumley 18hp 84, 86
1920 Bates Steel Mule half track tractor 22
1920 Keck-Gonnerman 18hp 85
1921 Huber Return Flue 22hp 40, 69
1921 LaFrance Brockway Torpedo Firetruck 155
1921 Ruggles Model 20 truck 110
1921 Titan 10-20hp 14, 18
1922 AB Farquhar model "K" 20hp steam traction engine 74, 85
1922 A.D. Baker 19-60hp 27, 28, 40, 69
1922 Minneapolis 17-30hp 29, 30
1923 American #207 4-0-4 53, 88
1923 Frick Eclipse Twin steam traction engine 68
1923 Huber return flue steam roller 65, 67, 69, 85
1924 Huber 25hp return flue steam traction engine. 56
1924 Universal #3-rock crusher 49
1925 Model "L" Rumley Oil Pull 15-25hp 5, 26
1925 Nichols and Shepard 25-50hp kerosene tractor 27, 28, 185
1925 Rumley Oil Pull Model L 15-25hp. 25
1926 Avery 12hp tractor, 39
1926 Eagle model "H" 20-40hp 78, 79, 188
1926 Fordson model F tractor 187
1926 Rumely Oil Pull model M 20-35 186
1926 Stillwell Coach 97
1927 Twin City. 17-28hp tractor 71, 70
1928 Baker 25-50hp 5, 14, 18, 20
1928 Cletrac 40hp 65
1928 Greyhound 25-52hp tractor, 32
1928 Hart Parr 28-50hp tractor 33
1928 "W" Rumley Oil Pull 20-30hp 18
1928 Rumley Oil Pull 25-40hp 14, 18, 20
1929 4-Pull tractor 65, 67
1929 Hart-Parr 28-50 188
1929 Paige automobile 49
1929 Rumely Oil Pull model Z 40-60 Super Power 190
1929 Wallis 12-20hp tractor 48
1930 Banting 20-35hp tractor 46
1930 Banting 20-40 191
1931 Harrison Jumbo 20hp 40, 69, 85
1933 Tructor tractor, 71
1935 Allis-Chalmers model E 25-40 191
1935 Parrett model 6 tractor 70

1935 Parrett tractor Model #6. 70
1936 Kaywood tractor 71, 72
1936 McCormick Deering WK40 tractor, 70
1937 Ford air-cooled experimental tractor 55
1937 Love tractor 80
1938 Worthington six cylinder 500hp 300 Kilowatt engine and generator set 68, 114
1939 Economy tractor 192
1939 Ford 9N 192
1940 Mayrath Standard 72
1941 McCormick Deering O-4 193
1942 Dodge Fire Truck 109
1945 Cletrac model HG crawler tractor 194
1945 Farmall model A 193
1948 Field Marshall tractor 81
1949 Centaur, 72
1949 Long model "A" 72
1951 Sheppard SD3 diesel tractor 195
1954 Chevy ½ ton pick truck 110
1966 Case model 1200 195
1972 replica ½-scale Case 65hp steam traction engine 27, 85

A

Advance-Rumley 20hp 26, 28
Aultman Taylor 30-60hp prairie tractor 49

C

Case 65hp steam traction engine 26, 27, 28, 40
Case 75hp 14, 18
Cooper-Bessemer model GAB 25hp stationary engine 78

F

Flour City 40-70 Special Road Tractors 146
Ford model T canopy Express 109

K

Keck-Gonnerman steam traction engine 183

L

Lauson 20-40hp 72

N

Nichols and Shepard 28, 69

P

Port Huron 19hp 28

R

Rumely Oil Pull model G 20-40, 185